T0336961

Mathematical Modelling in Biomedicine

Mathematics and Its Applications

Managing Editor:

M. HAZEWINKEL

Centre for Mathematics and Computer Science, Amsterdam, The Netherlands

Editorial Board:

F. CALOGERO, *Universita degli Studi di Roma, Italy*
Yu. I. MANIN, *Steklov Institute of Mathematics, Moscow, U.S.S.R.*
A. H. G. RINNOOY KAN, *Erasmus University, Rotterdam, The Netherlands*
G.-C. ROTA, *M.I.T., Cambridge, Mass., U.S.A.*

Y. Cherruault
Medimat, Université Paris VI, Paris, France

Mathematical Modelling in Biomedicine

Optimal Control of Biomedical Systems

D. Reidel Publishing Company

A MEMBER OF THE KLUWER ACADEMIC PUBLISHERS GROUP

Dordrecht / Boston / Lancaster / Tokyo

Library of Congress Cataloging in Publication Data

CIP

Cherruault, Y.
    Mathematical modelling in biomedicine.

    (Mathematics and its applications)
    Bibliography: p.
    Includes index.
    1.   Medicine–Mathematical models.   2.   Biology–Mathematical
models.   I.   Title.   II.   Series: Mathematics and its applications (D.
Reidel Publishing Company [DNLM: 1.   Biometry.   2.   Models,
Biological. QH 323.5 C522mb]
R853.M3C44   1985   574′.0724   85-25656
ISBN 90-277-2149-1

Published by D. Reidel Publishing Company
P.O. Box 17, 3300 AA Dordrecht, Holland

Sold and distributed in the U.S.A. and Canada
by Kluwer Academic Publishers,
190 Old Derby Street, Hingham, MA 02043, U.S.A.

In all other countries, sold and distributed
by Kluwer Academic Publishers Group,
P.O. Box 322, 3300 AH Dordrecht, Holland

Printed in The Netherlands

To Annick and Yann
for their contribution to my
inspiration and for their constant
encouragement and understanding
during the preparation of this book.

In Memory of my father

To my mother

CONTENTS

SERIES EDITOR'S PREFACE

Approach your problems from the right end and begin with the answers. Then one day, perhaps you will find the final question.

'The Hermit Clad in Crane Feathers' in R. van Gulik's *The Chinese Maze Murders*.

It isn't that they can't see the solution. It is that they can't see the problem.

G.K. Chesterton. *The Scandal of Father Brown* 'The point of a Pin'.

Growing specialization and diversification have brought a host of monographs and textbooks on increasingly specialized topics. However, the "tree" of knowledge of mathematics and related fields does not grow only by putting forth new branches. It also happens, quite often in fact, that branches which were thought to be completely disparate are suddenly seen to be related.

Further, the kind and level of sophistication of mathematics applied in various sciences has changed drastically in recent years: measure theory is used (non-trivially) in regional and theoretical economics; algebraic geometry interacts with physics; the Minkowsky lemma, coding theory and the structure of water meet one another in packing and covering theory; quantum fields, crystal defects and mathematical programming profit from homotopy theory; Lie algebras are relevant to filtering; and prediction and electrical engineering can use Stein spaces. And in addition to this there are such new emerging subdisciplines as "experimental mathematics", "CFD", "completely integrable systems", "chaos, synergetics and large-scale order", which are almost impossible to fit into the existing classification schemes. They draw upon widely different sections of mathematics. This programme, Mathematics and Its Applications, is devoted to new emerging (sub)disciplines and to such (new) interrelations as exempla gratia:

- a central concept which plays an important role in several different mathematical and/or scientific specialized areas;
- new applications of the results and ideas from one area of scientific endeavour into another;

xi

influences which the results, problems and concepts of one field of enquiry have and have had on the development of another.

The Mathematics and Its Applications programme tries to make available a careful selection of books which fit the philosophy outlined above. With such books, which are stimulating rather than definitive, intriguing rather than encyclopaedic, we hope to contribute something towards better communication among the practitioners in diversified fields.

At the moment there are many (relatively)) new areas where mathematical ideas and techniques are vigorously applied. Many of these are among the faster growing parts of mathematics. Mathematical biology and medicine (including pharmacology), or perhaps better, mathematical modeling in these areas, and the subsequent use of these models, could very well be the fastest growing ones of them all.

It does not sound too difficult: building a mathematical model of something. In reality it involves quite a lot, and there are aspects of this activity which smack of art rather than science. In particular it involves the matter of a scientist from the field in question and a mathematician getting to understand each other, a process which takes considerable amounts of time and effort and which has been described as painful.

It certainly helps in such a case to have a book available written by someone who has been through the mill several times, who has given lots of thought to the particular difficulties and pitfalls involved and who can write about it clearly.

Such are the contents of this book.

In addition biomedical mathematical models tend to generate different mathematical problems and questions compared to the more traditional physical and mechanical models, thus giving rise to new mathematical challenges. That is also what this book is about.

The unreasonable effectiveness of mathematics in science ...

Eugene Wigner

Well, if you know of a better 'ole, go to it.

Bruce Bairnsfather

What is now proved was once only imagined.

William Blake

As long as algebra and geometry proceeded along separate paths, their advance was slow and their applications limited.

But when these sciences joined company they drew from each other fresh vitality and thence forward marched on at a rapid pace towards perfection.

Joseph Louis Lagrange.

Bussum, August 1985   Michiel Hazewinkel

> *"Là où est l'amour des humains,*
> *là est aussi l'amour du métier."*
> HYPPOCRATE
> Préceptes, tome IX, Ed. Littré

PREFACE

This book would never have been published without the help
of the open-minded and benevolent friends who have contribu-
ted to our research programme in many ways, including finan-
cing the work. Most French research organisations have not
yet recognised the relevance of biomathematics, a multi-
disciplinary field. We have developed this field without
their help.

The author wishes to acknowledge the help of:

- M.G.E.N. (Mr. Pelinard and Mr. Lajeunie), which has
  financed our work for several years.

- Some pharmaceutical laboratories, specifically Sandoz and
  Servier (Dr. Kiechel, Miss Lavene, Mrs. Guerret,
  Dr. Poirier, Dr. J. F. Prost) that have signed contracts
  with the Medimat laboratory in order to develop the field
  of pharmacological mathematics.

- A number of firms (Specifically Bics (Mr. Dain), Worms bank
  (Mr. Janssen), B.N.P. (Mr. Simonpieri), F.N.A.C. (Mr. Essel,
  Mr. Kerinec), Diners Club (Mr. Gautier)) who were
  instrumental in realising our benefits from apprenticeship
  taxes.

- The Pierre and Marie Curie University (Paris VI), which
  has always fostered our field (Prof. Dry, Prof. Astier,
  Prof. Garnier, Prof. Proteau).

- Ligue Française contre la Vivissection (Mr. Duranton de
  Magny, Dr. Kalmar) for its financial assistance in our
  research.

- D. Reidel Publishing Company (Dr. Larner) and Prof. Hazewinkel (Centre for Mathematics and Computer Studies,
  Amsterdam), who asked me to write this book as they were
  interested in the field of biomathematics.

- Annick Cherruault for her essential bibliographical
  research work and her share in our work.

We thank all these people and companies, among others.
We owe them an important debt with respect to our scientific
work. We hope that our work will be developed further by
the young open-minded scientists interested in our book.

Paris, 1985   Y. CHERRUAULT

> *"D'abord, une grande racine de foi,*
> *ensuite, une grande boule de doute,*
> *enfin, une grande fermeté de dessein."*
> HAKUIN
> Maître ZEN (1685-1768)

Yves Cherruault is one of the founders of Biomathematics.
This book contains important new developments in this field,
introduces new ideas, and opens up new areas of research.

Yves Cherruault has succeeded in combining theory and practice to develop methods and procedures which connect mathematics to biology, medicine, and pharmacology. These inter-
relationships are illustrated by means of practical examples,
coming from the author's experience as head of the Medimat
laboratory, where these problems are approached and treated
both from a theoretical viewpoint and by means of practical
applications.

The problems considered are deterministic or statistical, and
it is necessary to increase the number of tools and techniques
available to scientists and engineers working in this field.

Chapter 1 of the book is an introduction. Here the author
presents general problems of modelling and various techniques.
He gives some examples and illustrates the difficulties
involved.

Chapters 2 to 4 chiefly concern pharmacological applications,
compartmental systems, dose-effects and concentration-effect
relationships. These topics provide all the information
necessary to define and solve the corresponding differential
systems, linear or not, with or without internal delays, and
their controls. Some new mathematical methods are developed,
such as the Alienor method, that enable one to find the
absolute extrema.

Therapeutical applications are covered in Chapter 5 and 6.
Chapter 5 covers the first modelling methods in medicine and
the treatment of hormonal equilibrium by Dr. Bernard Weil's
equations, the optimal control of this system is treated by
an original mathematical method. The regulation of glucose
using insulin is presented in Chapter 6.

Chapters 7 to 9, in which the more classical problems are
examined, introduce many novel approaches.
Chapters 7 and 8 illustrate some applications of integral
equations in biomedicine, in particular, their use in the

problems of catheter utilisation, knee articulation and diffusion of oxygen from the kidneys. New and powerful methods for defining and solving integral equations or mathematical systems are introduced, such as Caron's method. These methods can be implemented on microcomputers. Chapter 9 treats partial differential equations and systems, which model Kernevez's problem for membranes, gas exchange during breathing and other related topics.

Chapter 10 presents the important problem of optimality in human physiology. Professor J. Brocas has provided useful information on minimum necessary energy consumption. Applications are presented; for example, the problem of thermo-regulation.

Chapter 11 treats errors in computation and modelling. These errors, arising from computers and measurements, can be estimated using the methods of Vignes and sensitivity theory.

Chapter 12 treats several open problems: biological delayed systems, retrocontrolled systems, the simultaneous action of two or more drugs, numerical techniques for global optimiz- ation, optimization of industrial processes, and optimality criteria in physiology. Case studies are described for some of these problems.

Chapter 13 describes the present situation in the field and looks into the future of biomathematics.

The nervous system is not explicitly treated, but the chapters on hormonology and pharmacology can be used by neurologists and psychiatrists.

The interest and importance given to biomathematics varies from country to country. This book will outline our methods and concepts for mathematicians, biologists, physicians, pharmacologists and engineers. Applied mathematical problems originating in the fields of medicine and biology not only lead to new applications of existing algorithms but also to the creation of new ones. These, in turn, lead to new models, because none are available in most cases.

Dr. A. Guillez, Chairman
Symposium of mathematical modelling in Medicine and Biology,
International Congress of Cybernetics, Namur, 1983

INTRODUCTION

The aim of this book is to present mathematical methods for
building models of biomedical systems. For this, many
mathematical and numerical techniques are necessary.
Biomedical examples showing the usefulness of mathematical
and numerical tools will be introduced. Much of the work in
biomathematics is published in specialised reviews but books
in this area are rather scarce [19], [49]. Of course this
science is strongly multidisciplinary. The mathematician,
the biologist or the physician cannot develop research
alone. Multidisciplinary teams are needed for
biomathematics to exist and expand. Abstraction is only
used to obtain effective knowledge of concrete phenomena.
As in mathematics and physics, biomathematics involves many
difficulties, both from the theoretical and practical points
of view, but these difficulties are of a different kind. In
fact, the work of the biomathematician consists in
formalizing biological systems using systems of equations.
We shall see in the following some practical methods of
mathematical formalization. A first question arises: is
such a representation useful? The answer is 'yes' only if
it allows the evolution of the system to be predicted with
good approximation without further experiment. More
precisely the usefulness of a mathematical model is assured
if the system can be acted on by the use of "control"
parameters. Optimization or optimal control of biological
phenomena becomes possible and gives, when applied to drugs
for example, the possibility of optimal therapeutics. This
step consisting of associating a model with a biological
system is often difficult, due to the lack of physical
knowledge about the phenomenon. Indeed, most of the time we
have only experimental data and some results from literature
at our disposal. Much bibliographic research and many
trials with different mathematical models are needed to
find a well adapted model. A first representation is rarely
convenient and modifications will be necessary to adapt the
mathematical formalism to the experimental data.
Intuitively it is clear that every phenomenon can be

1

modelled but in practice we may meet some insurmountable
difficulties in building or managing the model. This will
be the case if the experimental data are insufficient or are
of low quality (poor approximation, for example). Likewise,
studying a too complex system can lead to a very complex
mathematical model which is beyond all known mathematical
and numerical methods. In this situation the solution
could be:

- develop new methods for studying the mathematical
  model
- propose a simplified model still in accordance with
  the numerical data
- renounce the study of the biological system because
  of insufficient knowledge or deficient mathematical
  and numerical techniques.

In the past, physical systems provided many problems
for mathematicians. They gave rise, in some cases, to new
theories and methods. Now, biological systems can give
interesting problems to mathematicians, perhaps involving
some new ideas, theories and methods. From a practical
point of view one can assert that mathematical models
idealize biological systems, but they give the possibility
of taking into account all the data about the phenomenon in
a concise and rigorous manner [22], [39].

The main mathematical methods involved are:

- data analysis for the determination of the
  fundamental variables
- analytical or numerical resolution of
  differential, partial differential and integral
  systems of equations
- parameter identification in models involving unknown
  functions
- optimization methods
- optimal control for finding optimal policies
  according to one or several criteria
- determination of optimal criteria allowing a unified
  view even when it is not realized from the
  available experimental data

In the following, all these ideas will be developed
using concrete examples.

CHAPTER 1

GENERAL REMARKS ON MODELLING

1.1 Definitions

As previously mentioned, a model is a system of mathematical equations (algebraic, differential, partial differential, integral) allowing all the known experimental data to be taken into account. A model must involve:

- the best possible understanding of the biological phenomenon
- the possibility of functioning (system optimization)

In the following we shall see how to distinguish between the state variables (describing the system evolution) and the control variables (acting on the system). Models are of two main types:

a)   knowledge models where mathematical equations are obtained as translations of the physical laws of the system, and
b)   simulation models giving a priori equations whose coefficients must be identified from available experimental data.

In practice we often build intermediate models using both physical laws and numerical experimental data. Although knowledge models are intellectually more satisfying, simulation models are equally useful. In the following the problem of model uniqueness will be considered. It is often a false problem because approximations (of a function, of a set of equations) are rarely unique. A model is, of course, an approximation of a biological phenomenon. Indeed, models differ from each other because they are based on different approximations or simplifications. In any case the final objective will involve the structure of the model. In practice one must choose the simplest model consistent with known experimental data and with permissible numerical errors. Simplicity is

3

characterised by the ease of solving the resulting
mathematical problems associated with the model. The main
constraint for a model must be its usefulness. Modelling
should not be a pretext for display of mathematical
expertise. It must involve new practical results and
justify hypotheses or intuitions. In biomathematics, as in
the concrete sciences, theory has to be controlled by
experience. Furthermore, it can be seen that modelling is
essential when complex biological systems are studied. For
example, the understanding of optimized or controlled
systems involves modelling. In the following we shall see
concrete models, but let us notice now that in general a
model can be represented schematically by:

- algebraic relations (linear or non-linear)
- differential, partial differential or integral
  equations

Finally, the use of approximation techniques [25] show
that every model can be brought back to a system of
algebraic equations. The general case is modelling by
algebraic relations.

1.2 The main techniques for modelling

1.2.1 COMPARTMENTAL ANALYSIS [35]

This method is the most popular in biomathematics. It is
also the simplest and leads quickly to mathematical
relationships. Compartmental modelling is very convenient
for biochemical transformations but can also be applied to
various problems. Firstly, the substance being studied is
divided into different compartments which are equivalent
classes from a mathematical point of view. These
compartments are defined from physical properties
(localisations, physical or electrical states, chemical
states...) Compartmental analysis consists of studying
the exchanges of matter between the compartments as a
function of time, t. To obtain mathematical equations some
supplementary hypothesis is needed to quantify the exchanges
of matter. For instance, let us suppose that the flow from
compartment i to compartment j is proportional to the
quantity $x_i(t)$ contained in the source compartment i. The
writing of the equations becomes almost obvious if we do a

mass balance in each compartment during a time interval dt.
Consider a compartmental system with n compartments, linked
in both directions, as shown in Figure 1.

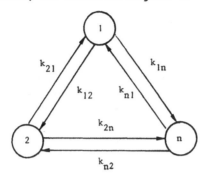

FIG. 1 .

It is a closed system. A mass balance involving the
following relation:

  instantaneous variation of quantity in compartment i
            = input flow - output flow

leads to the following differential equation:

$$\dot{x}_i = \sum_{\substack{j=1 \\ j\neq 1}}^{n} k_{ji}x_j - \sum_{\substack{j=1 \\ j\neq 1}}^{n} k_{ij}x_i \qquad (1.1)$$

which is valid for i = 1, 2, . . , n

In (1.1) the k_{ij} are the positive proportionality parameters
coming from our hypothesis. Generally the initial
conditions are known:

$$x_i(0) = \alpha_i \qquad \text{(known)} \qquad (1.2)$$

Setting:

$$k_{ii} = - \sum_{\substack{j=1 \\ j\neq 1}}^{n} k_{ij} \qquad (\leqslant 0)$$

we obtain the following simplified differential system:

$$\dot{x}_i = \sum_{j=1}^{n} k_{ji} x_j, \quad i = 1, \ldots, n$$

$$x_i(0) = \alpha_i \tag{1.3}$$

which is linear in the $x_j(t)$. Such a system is not difficult to solve if we know the coefficients k_{ij} and α_i [24], [38]. It is possible to use either explicit methods, if n is small, or numerical techniques where n is large. (Euler, Runge-Kutta techniques . . .). Unfortunately, as we shall see later, the main difficulty for system (1.3) is not its resolution but the evaluation of the unknown parameters k_{ij} using experimental observations which can be very poor.

A system as in (1.3) is based on a linear hypothesis. We can use other hypotheses and then adapt a compartmental model to each concrete biological phenomenon. One can, for example, suppose that the flow from i to j is proportional to the product of the quantities in the i and j compartments. System (1.3) then becomes:

$$\dot{x}_i = \sum_{\substack{i=1 \\ j \neq i}}^{n} k_{ji} x_i x_j - \sum_{\substack{i=1 \\ j \neq i}}^{n} k_{ij} x_i x_j \tag{1.4}$$

$$x_i(0) = \alpha_i$$

and we obtain a non-linear differential system.

Another possible hypothesis (there are many other possibilities which can be adapted to each concrete case) consists in translating the properties of enzymatic systems. Let us consider the following compartmental model [72], in which k_m, V_m characterises an enzymatic relation.

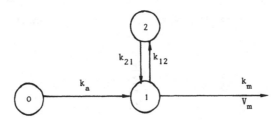

FIG. 2 .

The equations of the model shown in Figure 2 are given by:

$$\dot{x}_0 = -k_a x_0$$

$$\dot{x}_1 = k_a x_0 - (k_{12} + (V_m/(k_m + x_1(t))))x_1(t) + k_{21}x_2(t)$$

$$\dot{x}_2 = k_{12}x_1 - k_{21}x_2$$

(1.5)

$$x_1(0) = x_{10}, \quad x_2(0) = x_{20}, \quad x_0(0) = D$$

We also obtain a non-linear differential system because of the term:

$$V_m/(k_m + x_1(t))$$

translating an enzymatic transformation.

In some cases it can be useful to consider compartmental models as in (1.3),(1.4) or (1.5) where the coefficients k_{ij} depend on the time t. We shall use these models in the following to study optimization of drug therapeutics. To summarize, it can be asserted that compartmental modelling can fit every biological system that depends on time t. A question arises: What kind of model have we realized? At first sight compartmental analysis looks like a knowledge model but it is often difficult to give a biological justification to the chosen hypothesis. In this case, the compartmental model can be classified as belonging to the set of intermediate models.

Remark 1. Compartmental models as in (1.3) have characteristic properties. In particular, if $k_{ij} \geqslant 0$ for $i \neq j$, and $k_{ii} \leqslant 0$, it follows that if $x_i(0) \geqslant 0$ for all i then the solution $x_i(t)$ is positive for all i and $t > 0$ [3a].

Remark 2. From a biological point of view it is not possible to admit systems such as (1.3) having positive characteristic values. The consequence would be that some $x_i(t) \to +\infty$ which is obviously impossible. In fact complex characteristic values are not excluded theoretically, but they are not generally admitted by biologists. So, in the following, we shall use another hypothesis: characteristic values of system (1.3) are distinct, negative or equal to zero.

1.2.2 SYSTEMS WITH DIFFUSION-CONVECTION REACTIONS

Many biological systems (respiration, exchanges between capillaries and tissues, artificial kidney, ...) involve physical laws such as those of diffusion and convection reactions. The associated models, based on the mathematical translation of physical properties, can be considered as knowledge modelling. For example, consider the respiratory system [11], [18], where three gases are taken into account (oxygen O_2, carbon dioxide CO_2 and nitrogen N_2). The problem consists in modelling the variations (in time and space) of these gases in the alveolus and in the blood. The alveolus is the unitary element of the lung. It can be represented schematically by a tube of length L, radius R and thickness e (Figure 3). The tube is open at x = 0 and closed at x = L.

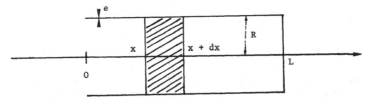

FIG. 3 .

The blood circulates around the alveolus. The main role of the alveolus is to ensure the gas exchange between the blood and the surrounding air. Blood needs oxygen for nourishing tissues and it is necessary to expel the carbon dioxide.

Let y(x,t), v(x,t), s(x,t) be the partial pressures of O_2, CO_2 and N_2 in the tube and z(x,t), w(x,t), u(x,t) be the partial pressures of the same gases in the blood.

The variation equations are obtained by doing a mass balance between x and x + dx during dt. Let us build the equation for O_2. The global variation of the quantity of oxygen during dt is equal to:

$$Sdx.y(x,t+dt) - Sdx.y(x,t) \simeq S.dxdt \, (\partial y/\partial t)$$

where $S = \pi R^2$

But this global variation is equal to the sum of elementary variations due to the phenomena being considered

- diffusion, convection and diffusion through the wall. If we neglect the radial diffusion, the longitudinal diffusion gives the following elementary variation in our small element (diffusion law [6]):

$$dt(-K.S.\partial y/\partial x(x,t) + K.S.\partial y/\partial x(x+dx,t)) \simeq K.S.\partial^2 y/\partial x^2 dxdt$$

where K is a known diffusion coefficient. The previous expression is obtained by the relation:

variation due to diffusion
= quantity at x - quantity at x + dx

Let $\gamma(x,t)$ be the convection flow for the gases. Then the elementary variation for oxygen according to the law of convection is given by:

$$y(x,t).\gamma(x,t)dt - y(x+dx,t)\gamma(x+dx,t)dt \simeq - \partial/\partial x(y.\gamma)dxdt$$

The calculation of the variation due to the diffusion through the wall remains. We apply Fick's law which gives:

$$- D_{O_2}(2\pi R/e)(y-z)dxdt$$

where D_{O_2} is a diffusion coefficient associated with O_2. This is the reaction part of our formula. The following differential equation is therefore obtained (after simplification by dx.dt):

$$\partial y/\partial t = K.\partial^2 y/\partial x^2 - (1/S)\partial/\partial x(y\gamma) - D_{O_2}(2/Re)(y - z)$$

The other two relations, for CO_2 and N_2 in the tube, are obtained in the same way.

For oxygen in the blood (variable z) there are some slight modifications which are described below. Let $Q(t)$ be the blood flow (supposedly independent of x, which is a good approximation). The elementary variations due to diffusion (ordinary and through the wall) are obtained as before. For convection, a solubility coefficient $\$_{O_2}$, must be introduced. Then the elementary variation is given by:

$$Q(t).\$_{O_2} z(x,t)dt - Q\$_{O_2} z(x+dx,t)dt \simeq - Q\$_{O_2} \partial z/\partial x \, dxdt$$

and the partial differential equation for z is the following:

$$\partial z/\partial t = K_S \partial^2 z/\partial x^2 - Q/S_1 \ s_{O_2} \partial z/\partial x - D_{O_2}(2\pi R/eS_1)(z-y)$$

where S_1 is the blood part of the equation, and is not usually equal to S.

Now it is possible to write the complete set of equations:

$$\partial y/\partial t = K\partial^2 y/\partial x^2 - 1/S\partial/\partial x(y\gamma) - (2D/Re)(y - z)$$

$$\partial v/\partial t = K\partial^2 v/\partial x^2 - 1/S\partial/\partial x(v\gamma) - (2D/Re)(v - w)$$

$$\partial s/\partial t = K\partial^2 s/\partial x^2 - 1/S\partial/\partial x(s\gamma) - (2D/Re)(s - u)$$

$$\partial z/\partial t = - (Q/S_1)s_{O_2} \partial z/\partial x - (2\pi RD/eS_1)(z - y)$$

$$\partial w/\partial t = - (Q/S_1)s_{CO_2} \partial w/\partial x - (2\pi RD/eS_1)(w - v)$$

$$\partial u/\partial t = - (Q/S_1)s_{N_2} \partial u/\partial x - (2\pi RD/eS_1)(u - s)$$

$$(1.6)$$

where v and s are partial pressures of CO_2 and N_2 in the tube and w and u are partial pressures of the same gases in the blood. In (1.6) we neglected the diffusion terms because K_S (for the three gases) is very small. It is possible to add the following relation:

$$y + v + s = P_B \qquad (1.7)$$

which means that the sum of partial pressures in the tube is equal to the atmospheric pressure P_B. In fact, the tube is open at x = 0. The model (1.6),(1.7) is obviously a knowledge model. Furthermore, we may remark that the relations are coupled (y,z), (v,w), (s,u) so it suffices to have a numerical method to solve one pair of equations. Later we shall see original numerical methods to solve (1.6) and (1.7) which are based on the utilization of a global optimization technique.

1.2.3. SIMULATION MODELS

When the available data is insufficient, a simulation model is the only one possible for studying a biological system. Initially it is necessary to define the main variables using, for example, data analysis techniques or the intuition or experience of the biologist. These variables are then related by a priori systems of mathematical relations. These mathematical systems may be:

(i) a linear differential system

$$\dot{x}_1 = a_{10} + a_{11}x_1(t) + \ldots + a_{1n}x_n(t) + b_1$$
$$\cdot$$
$$\cdot$$
$$\cdot \qquad\qquad\qquad\qquad\qquad\qquad\qquad (1.8)$$
$$\cdot$$
$$\dot{x}_n = a_{n0} + a_{n1}x_1(t) + \ldots + a_{nn}x_n(t) + b_n$$

with initial conditions which may be known or unknown.

The x_i's are the variables playing an important role in the biological system. They may be quantities, concentrations, partial pressures, . . . In (1.8) the coefficients a_{ij}, b_i are unknown and must be calculated from experimental data on $x_i(t)$.

For instance $x_1(t), \ldots, x_n(t)$ may be measured for $t = t_j$, $j=1, \ldots, m$. A more complicated problem consists in identifying the a_{ij} and b_i when only $x_1(t), \ldots, x_p(t)$ is measured ($p \leqslant n$) for $t = t_j$, $j= 1, \ldots, m$. In this case uniqueness of identification is not always ensured even when general theorems on uniqueness and existence are used.

(ii) If the biological phenomenon depends only on time but is non-linear the following model can be tried:

$$\dot{x}_i = f_i(x_1, x_2, \ldots, x_n t), \qquad\qquad i=1, \ldots, n \qquad (1.9)$$

In this case the functions f_i are to be identified from experimental data. To simplify, f_i may be chosen as a polynomial and therefore polynomial parameters are to be identified. Even with this simplification the identification problem may become insoluble because of the large number of parameters to be identified. In practice, optimization methods are used to solve this type of identification problem leading to non-linear differential models.

(iii) Other mathematical equations are also possible. For instance, partial differential equations when the phenomenon depends both on time and space:

$$\partial x_i / \partial t = a_i \partial x_i / \partial x + F_i(x_1, \ldots, x_n), \qquad i=1, \ldots, n \quad (1.10)$$

(iv) Integral equations are also possible simulation models. If the phenomenon depends not only on past and present but also on the future then integral relations of Fredholm type [69] are convenient. For example:

$$x_i(t) = \int_0^T f_i(x_1(u), \ldots, x_n(u), t) \, du, \quad i=1, \ldots, n \quad (1.11)$$

In (1.11) the functions f_i have been identified from some experimental data. Of course it can be a difficult problem from the theoretical and numerical point of view.

(v) The simplest way is to relate the main variables by an algebraic system (linear or non-linear). For instance:

$$x_i = f_i(x_1, \ldots, x_n), \qquad i=1, \ldots, n \quad (1.12)$$

where f_i is a polynomial function of degree m. In some cases, one equation as in (1.12) is sufficient if we need only to relate one variable to another. As previously, the functions f_i have to be identified from experimental data using an optimization method. These techniques will be developed in the next chapters which deal with concrete models.

1.3 Difficulties in modelling

Much time must be spent in obtaining a good understanding of the biological phenomenon. Bibliographic research is necessary and is often long and difficult. Knowledge is to be found in numberless publications. The determination of physical laws is generally not obvious and requires a great deal of multi-disciplinary discussion. The objectives of the model must be clarified because they involve the structure of the mathematical equations and their complexity. From a mass of data and parameters it is necessary to choose the most important variables and coefficients. If we intend to build a knowledge model the relevant physical laws will be used. Most of the equations

of the model are obtained by writing the conservation laws
(of mass, energy, motion, electricity, . . .). To be
successful we must make as many simplifying hypotheses as
are needed to obtain a useful model. Of course, these
hypotheses have to be reasonable from a biological point of
view: the approximations must be justified. Simplification
is necessary because an unduly complex model (mathematically
speaking) cannot be developed. If necessary a simple model
may always be extended to take account of new results or new
experiences.

There are also some difficulties when using
mathematical, numerical or informational methods associated
with a mathematical model. Firstly, the analytical
resolution of the equations of the model is rarely possible.
It is often necessary to use numerical methods and sometimes
to invent new techniques because existing methods are not
satisfactory. Existence and uniqueness of solutions is
always difficult to prove when considering practical models.
The identification of unknown parameters and functions
constitutes a real and difficult problem in modelling. We
do not ignore general theorems and each concrete system must
be studied using our intuition and imagination. Even in
simple models, such as compartmental ones, there are
unsolved identification problems. Classical analysis fails
and therefore many problems arise and remain open. In
practice, even when there is not uniqueness, a model must be
chosen, so additional choice criteria have to be defined
according to the biological properties. As said,
previously, identification problems lead to optimization
methods and in spite of a rich literature on this subject
many difficulties remain. In particular, when we are
looking for an optimum (maximum or minimum) almost all
techniques converge on a local optimum. In our
identification or optimization problems it is important to
have global optimization techniques at our disposal. In the
following some optimal control problems will arise. They
are particular cases of optimization problems but are
especially difficult to solve (mathematically and
numerically). The greater part of this book will be related
to optimization theory. It is a crucial theory for
biomathematics. The last question to arise is: what can be
modelled in biology and medicine? Theoretically every
biological phenomenon may be interpreted by a mathematical
system. But the efficiency is conditioned by the quality
and quantity of experimental data. Without good data, the

probability of obtaining an effective model is almost nil.
We now have to consider modelling and its difficulties in
concrete biological situations.

CHAPTER 2

IDENTIFICATION AND CONTROL IN LINEAR COMPARTMENTAL ANALYSIS

2.1 The identification problem

Let us recall the general equations for a linear compartmental model:

$$\dot{x}_i = \sum k_{ji} x_j, \qquad i = 1, \ldots, n$$
$$x_i(0) = \alpha_i \tag{2.1}$$

which are associated with n compartments, all connected in both directions. These compartmental models are very useful for studying drug kinetics in human or animal organisms. For example, compartmental models such as the following are relevant to beta-blockers, which are drugs that act on the blood pressure and heart rate.

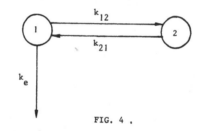

FIG. 4 .

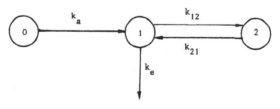

FIG. 5 .

Figure 4 corresponds to intravenous injection of the drug, and Figure 5 to oral absorption. 0 is the absorption

compartment (gastro-intestinal tract), 1 is the blood
compartment and 2 is the "deep action" compartment of the
drug. Before solving the general identification problem,
let us consider the cases represented by Figure 4 and Figure
5. The compartmental model shown in Figure 4 is translated
by [4], [27], [40].

$$\dot{x}_i = - (k_{12}+k_e)x_1 + k_{21}x_2$$
$$\dot{x}_2 = k_{12}x_1 - k_{21}x_2$$

(2.2)

with known initial conditions:

$$x_1(0) = \alpha \qquad , \qquad x_2(0) = 0 \qquad (2.3)$$

In (2.2) and (2.3) the $x_i(t)$ are the quantities of drug
at time t in compartment i. The initial conditions (2.3)
mean that a quantity has been given (in an instantaneous
manner) at time t = 0, in compartment 1. We suppose that at
this time there is no drug in the second compartment. In
practice, only $x_1(t)$ is measured for t = t_j, where
j = 1, . . ,m. The identification problem is as follows:

Find k_{12}, k_{21}, k_e knowing $x_1(t_j)$, j = 1, . . ,m where
k_{12}, k_{21}, k_e, x_1 and x_2 are solutions of (2.2), (2.3).

Thus, the problem of existence and uniqueness of k_{ij} is
posed. First we note that x_1, x_2 are linear combinations of
$\exp(\lambda_1 t)$, $\exp(\lambda_2)$ where λ_1 and λ_2 are the characteristic
values of the differential system (2.2). We suppose these
characteristic values are real and negative. These
properties arise from biochemical considerations [72].

Therefore:

$$x_1 = \beta_1^1 \exp(\lambda_1 t) + \beta_1^2 \exp(\lambda_2 t)$$
$$x_2 = \beta_2^1 \exp(\lambda_1 t) + \beta_2^2 \exp(\lambda_2 t)$$

(2.4)

Putting (2.4) into (2.2) and identifying the coefficients of
$\exp(\lambda_1 t)$ and $\exp(\lambda_2 t)$ involves the following algebraic
relations:

$$\beta_1^1 \lambda_1 = - (k_{12} + k_e)\beta_1^1 + k_{21}\beta_2^1$$

$$\beta_1^2 \lambda_2 = - (k_{12} + k_e)\beta_1^2 + k_{21}\beta_2^2$$

$$\beta_2^1 \lambda_1 = k_{12}\beta_1^1 - k_{21}\beta_2^1 \qquad (2.5)$$

$$\beta_2^2 \lambda_2 = k_{12}\beta_1^2 - k_{21}\beta_2^2$$

The initial conditions (2.3) give the two relations:

$$\beta_1^1 + \beta_1^2 = \alpha \ , \qquad \beta_2^1 + \beta_2^2 = 0 \qquad (2.6)$$

We cannot obtain any other relations between our unknown k_{12} ,k_{21}, k_e. We must now use our data to solve the identification problem. The measured $x_1(t_j)$ allow the determination of:

$$\beta_1^1, \ \beta_1^2, \ \lambda_1, \ \lambda_2$$

For this we associate the positive functional:

$$J = \sum_{j=1}^{m} [x_1(t_j) - \sum_{i=1}^{2} \beta_1^i \exp(\lambda_i t_j)]^2 \qquad (2.7)$$

J depends on β_1^1, β_1^2, λ_1, λ_2 and it is possible to look for its minimum as a function of these parameters, that is:

$$\underset{\beta_1^1, \ \beta_1^2, \ \lambda_1, \ \lambda_2}{\text{Min J}} = J(\beta_1^{1*}, \ \beta_1^{2*}, \ \lambda_1^*, \ \lambda_2^*) \qquad (2.8)$$

Later, we shall describe some optimization techniques to solve (2.8) numerically, but we can already remark that if J = 0 for the optimum values, then:

$$x_1(t_j) = \sum_{i=1}^{2} \beta_1^i \exp(\lambda_i t_j) \quad \text{for } j = 1, \ \ldots \ ,m$$

and thus $x_1(t)$ is consistent with a two-compartment model. In practice, it suffices to satisfy $J \simeq 0$, that is $J = \epsilon$, where the approximation ϵ depends on the errors in the $x_1(t_j)$.

<u>Remark</u>   The above technique for determining β_1^1, β_1^2, λ_1, λ_2 gives a way of knowing whether or not some data are consistent with a compartmental model. In fact, having $x_j(t)$ for $t = t_j$, $j = 1, \ldots, m$, $x_i(t)$ will be consistent with an n-compartment model if we can find:

$$\beta_i^1, \beta_i^2, \ldots, \beta_i^n, \lambda_1, \lambda_2, \ldots, \lambda_n$$

such that:

$$J_i = \sum_{j=1}^{m} [x_i(t_j) - \sum_{s=1}^{n} \beta_i^s \exp(\lambda_s t_j)]^2$$

allows a minimum for which J is almost equal to zero.
   If we have p data:

$$x_1(t), \ldots, x_p(t) \text{ measured for } t = t_j, \ j = 1, \ldots, m$$

and then we introduce p functionals:

$$J_1, \ldots, J_p$$

that must be minimized as functions of the β_i^s and λ_s (the λ_s have to be the same for all the $x_1, \ldots, x_p$ and λ_s is independent of i), and if every function J_i is near zero at the minimum, our p data, $x_1, \ldots, x_p$ are compatible with an n-compartment model.
   The identification problem becomes easy to solve. Indeed (2.5), (2.6) is an algebraic system of five independent equations and five unknowns:

$$\beta_2^1, \beta_2^2, k_{12}, k_{21}, k_e$$

The system is as follows:

$$\beta_1^1 \lambda_1 = -(k_{12} + k_e)\beta_1^1 + k_{21}\beta_2^1$$

$$\beta_1^2 \lambda_2 = -(k_{12} + k_e)\beta_1^2 + k_{21}\beta_2^2$$

$$\beta_2^1 \lambda_1 = \quad k_{12}\beta_1^1 \quad\quad - k_{21}\beta_2^1 \qquad\qquad (2.9)$$

$$\beta_2^2 \lambda_2 = \quad k_{12}\beta_1^2 \quad\quad - k_{21}\beta_2^2$$

$$\beta_2^1 + \beta_2^2 = \quad 0$$

(2.9) is a non-linear algebraic system which can be reduced to a <u>linear</u> system. In fact, $k_{21}\beta_2^1$ and $k_{21}\beta_2^2$ may be obtained from the third and fourth equations in (2.9) and substituted in the first two equations. The fourth equation when subtracted from the third gives yet another linear equation. We obtain four linear equations with four unknowns k_{12}, k_e, β_2^1, β_2^2. The determinant can be calculated and, if it is not equal to zero, we obtain a unique solution for k_{12}, k_e, β_2^1, β_2^2. The value of k_{21} can be calculated from one of the first four equations of (2.9). Unfortunately this method of linearizing the initial non-linear equations cannot be generalized. The general case will be considered in the following. Let us come back to the compartmental model shown in Figure 5. The mathematical equations are:

$$\dot{x}_0 = - k_a x_0$$

$$\dot{x}_1 = - (k_{12} + k_e)x_1 + k_{21}x_2 + k_a x_0$$

$$\dot{x}_2 = \quad k_{12}x_1 - k_{21}x_2$$ (2.10)

$$x_1(0) = x_2(0) = 0 \quad , \quad x_0(0) = D$$

and the identification problem consists of finding k_a, k_{12}, k_e, k_{21} knowing $x_1(t_j)$ for $j = 1, \ldots, m$.

As before, the knowledge of $x_1(t_j)$ gives the characteristic values $\lambda_1, \lambda_2, \lambda_3$ and the $\beta_1^1, \beta_1^2, \beta_1^3$. Note the structure of (2.10) involves the relation that $-k_a$ is equal to one of the λ_i, but we do not know which one. The first equation gives:

$$x_0 = D \cdot \exp(-k_a t).$$

We may use the same technique as for the 2-compartment model. It works well and also leads to a linear compartmental model by modification of non-linear initial relations obtained by identifying the coefficients of $\exp(\lambda_1 t)$, $\exp(\lambda_2 t)$, $\exp(\lambda_3 t)$.

Here we present another general technique for obtaining all the relations between the coefficients k_{ij}. Firstly, we note that the data $x_1(t_j)$, $j = 1, \ldots, m$ completely determines $x_1(t)$ for all t because, as previously said, we can identify the λ_i and β_1^i of:

$$x_1(t) = \beta_1^1 \exp(\lambda_1 t) + \beta_1^2 \exp(\lambda_2 t) + \beta_1^3 \exp(\lambda_3 t)$$

by an optimization technique. To do this we introduce the positive functional:

$$J_1 = \sum_{j=1}^{m} [x_1(t_j) - \sum_{i=1}^{3} \beta_1^i \exp(\lambda_i t_j)]^2$$

which is minimised as a function of the λ_i and β_1^i. Then we introduce linear combinations of $x_i^{(n)}(0)$ so as to obtain combinations of the k_{ij}. For example:

$$\dot{x}_1(0) = k_a x_0(0) = k_a D$$

which involves the linear relation:

$$\lambda_1 \beta_1^1 + \lambda_2 \beta_1^2 + \lambda_3 \beta_1^3 = k_a D$$

which gives the value of k_a.

Then:

$$\ddot{x}_1(0) = - (k_{12} + k_e) k_a D = \beta_1^1 \lambda_1^2 + \beta_1^2 \lambda_2^2 + \beta_1^3 \lambda_3^2$$

gives us $k_{12} + k_e$ as a function of known quantities. Furthermore:

$$\dddot{x}_1(0) = - (k_{12} + k_e)[- (k_{12} + k_e) k_a D + k_a D] + k_a \ddot{x}_0(0)$$
$$+ k_{21} k_{12} \dot{x}_1(0)$$

allows $k_{21} k_{12}$ to be obtained as a function of known quantities. Let us continue and calculate:

$$x_1^{(4)}(0) = - (k_{12} + k_e)\dddot{x}_1(0) + k_a \dddot{x}_0(0)$$
$$+ k_{21} \dddot{x}_2(0)$$
$$= \quad \ldots \ldots \ldots \ldots \ldots$$
$$+ k_{21}[k_{12}\ddot{x}_1(0) - k_{21}\ddot{x}_2(0)]$$
$$= \quad \ldots \ldots \ldots \ldots \ldots$$
$$+ k_{21} k_{12} \ddot{x}_1(0) - k_{21}^2 k_{12} \dot{x}_1(0) + k_{21}^3 \dot{x}_2(0)$$

This equation gives $k_{21}^2 k_{12}$ as a function of known quantities.

To sum up, we have calculated, k_a, $k_{12} + k_e$, $k_{12}k_{21}$ and $k_{12}k_{21}^2$. Values of k_a, k_{12}, k_{21} and k_e are easily obtained, but the relations given by this general technique are not linear. The extension to a general n-compartment system is obvious, but the relations between the k_{ij} may be difficult to write and to solve.

Another general method exists which is based on the Laplace transform [4], [42]. Let us first apply it to system (2.2) which is represented in Figure 4. Taking the Laplace transform of (2.2) gives a functional system without derivatives. Where s is the new variable associated with the Laplace transform, we have:

$$s\hat{x}_1(s) - \alpha = - (k_{12} + k_e)\hat{x}_1(s) + k_{21}\hat{x}_2(s)$$
$$s\hat{x}_2 \quad = \quad k_{12}\hat{x}_1(s) - k_{21}\hat{x}_2 \tag{2.11}$$

where $\hat{x}_i$ designates the Laplace transform of x_i.

The objective is to obtain a relation between the known functions or coefficients. Here we measured $x_1(t)$ and so we want to find a relation between α and $x_1(s)$. To do this it is necessary to eliminate x_2 using the second equation in (2.11).
From:

$$\hat{x}_2 = k_{12}\hat{x}_1 / (s + k_{21})$$

the following relation:

$$(s + k_{12} + k_e)\hat{x}_1 - \alpha = k_{21}k_{12}\hat{x}_1/(s + k_{21})$$

is obtained. In other words we have:

$$\hat{x}_1/\alpha = (s + k_{21})/[(s + k_{12} + k_e)(s + k_{12}) - k_{12}k_{21}]$$

Then we can say that the fractional rational function of the second member is known because $x_1(t)$ and α are known and therefore $\hat{x}_1(s)$ and α, and the numerical coefficients of this fraction are computable.

For example, we can introduce the functional:

$$J = \sum_{j=1}^{m} [\hat{x}_1(s_j)/\alpha - (s_j + A)/(s_j^2 + Bs_j + C)]^2$$

which is minimized as a function of A, B, C. In fact this minimization can be reduced to a linear problem.   It suffices to introduce the new functional:

$$J_1 = \sum_{j=1}^{m} [(\hat{x}_1(s_j)/\alpha) \cdot (s_j^2 + Bs_j + C) - (s_j + A)]^2$$

Min J_1 involves $\partial J_1/\partial A = \partial J_1/\partial B = \partial J_1/\partial C = 0$
A,B,C

(necessary conditions for an optimum).

The last three relations are linear in A, B, C because J_1 is a second-degree polynomial in A, B, C. Therefore A, B, C are obtained by solving a linear algebraic system. Returning to our own identification problem, the following relations are obtained:

$$k_{12} = A$$

$$k_{12} + k_{21} + k_e = B \qquad\qquad (2.12)$$

$$k_e k_{21} = C$$

The k_{ij}'s are uniquely determined using (2.12), but note that our algebraic system is non-linear even with only two compartments. Let us now considere the general case where the compartmental system is represented by the relations:

$$\dot{x} = Ax + u$$
$$\qquad\qquad (2.13)$$
$$x(0) = 0$$

where A is the matrix of the k_{ij}'s, and u an input function.
The observations from experimental data are:

$$z = B \cdot x$$

where A, B are matrices but only B is known, and x and u are n-vectors.

In the first case (Figure 4), $u = (\alpha\delta_{(0)}, 0, \ldots, 0)$ where $\delta_{(0)}$ is the Dirac delta-function at time 0. In the second case (Figure 5),

$$u(t) = (D\delta_{(0)}, 0, \ldots, 0)$$

The Laplace technique consists in looking for a relation between the known functions $\hat{u}$ and $\hat{z}$. The previous developments are easy to generalize. Indeed (2.13) becomes:

$$s\hat{x} = A\hat{x} + \hat{u} \quad \text{or} \quad \hat{x} = (sI-A)^{-1}\hat{u}$$

if (sI-A) has an inverse.

From $\quad \hat{z} = B\hat{x} \quad$ we obtain:

$$\hat{z} = B(sI-A)^{-1}\hat{u} \tag{2.14}$$

As can be easily proved $B(sI-A)^{-1}$ is a matrix whose terms are rational functions of s.

(2.14) gives all possible relations between the coefficients k_{ij} of matrix A. These relations are strongly non-linear. If A is (n x n) then the polynomial relations associated with (2.14) may be of degree n. This was the case in our particular example - a second-degree system for a two-compartment model. The general method for identifying an n-compartment model using the first technique based on the particular expression of $x_i(t)$ remains. We shall prove the following theorem [11], [12].

Theorem 2.1.1 Suppose that the characteristic values $\lambda_1, \cdots, \lambda_n$ of A (in $\dot{x} = Ax$) are distinct and negative. Suppose also, that the components z_i of the observation vector z in z = Bx can be uniquely expressed in the form:

$$z_i = \sum_{k=1}^{n} c_k^i \exp(\lambda_k t)$$

where the c_k^i and λ_k are known, or if necessary determined by numerical methods. Then if there exists a non-singular matrix T, such that:

$$BT = \Lambda_1 \quad \text{where} \quad \Lambda_1 = (c_k^i) \quad \text{and} \quad T.1 = \alpha$$

with $\alpha = (\alpha_i)$ $(x_i(0) = \alpha_i, i = 1, \cdots, n)$, the matrix A from $x = Ax$, $x(0) = \alpha$ (α given), can be identified. Furthermore if T is unique then A is uniquely identifiable.

The proof is very simple. Introduce Λ, the diagonal matrix formed with the proper values λ_i of A ($\Lambda = (\lambda_i)$) and the differential system $\dot{y} = \Lambda y$, $y(0) = 1$ whose solutions are $y_k = \exp(\lambda_k t)$. Set $x = Ty$, where T is the transition matrix from y, which is known, to x, which is partially unknown. Furthermore, note that $z = \Lambda_1 y$. Since $x = Ty$, we obtain:

$$\dot{x} = T\dot{y} = T\Lambda y = T\Lambda T^{-1} x$$

and

$$x(0) = T.y(0) = T.1 = \alpha$$

On the other hand we have:

$$z = \Lambda_1 y = \Lambda_1 T^{-1} x = Bx$$

We conclude that $T\Lambda T^{-1}$ is a solution of our identification problem. Indeed if we set $A = T\Lambda T^{-1}$ it follows that $\dot{x} = Ax$ and $x(0) = \alpha$ and $z = Bx$. A satisfies the differential system $\dot{x} = Ax$, $x(0) = \alpha$ and also $z = Bx$, representing the observations. It is easy to show that uniqueness of T implies uniqueness of k_{ij} (or A).

Remark If $B = 0$, that is nothing is observed about our model, then every singular matrix T satisfying $T.1 = \alpha$ is acceptable. Conversely, if $B = I$ (all $x_i(t)$ are known) or more generally if B is non-singular, then T is unique and equal to Λ_1. Alternatively, one can say that non-singularity of Λ_1 is equivalent to the existence and uniqueness of matrix T. Therefore, in practice, this theorem is not very interesting because is implies uniqueness only when all $x_i(t)$, $i = 1, \cdot \cdot , n$ are known. However, for the proof we did not use all the properties of the compartmental system. For example, matrix A contains some zeros, especially when there is no exchange between i and j ($k_{ij} = 0$), and some additional relations. In the system described in Figure 4 we have $\dot{x}_1 + \dot{x}_2 = -k_e x_1$. In the system in Figure 5, we have $\dot{x}_0 + \dot{x}_1 + \dot{x}_2 = -k_e x_1$. These supplementary relations must be taken into account when defining the matrix T. They are not always easy to express mathematically. No general theorem has been proved which includes the particular features of a compartmental system. The problem remains open for applied mathematicians [29].

In practice, one may build the whole set of mathematical relations between the unknowns. Notice that:

$$x_i(t) = \sum_{k=1}^{n} \beta_i^k \exp(\lambda_k t) \quad \text{for an n-compartment system.}$$

Putting these expressions in the differential formula:

$$\dot{x}_i = \sum_{j=1}^{n} k_{ji} x_j \qquad i = 1, \ldots, n$$

and identifying the expressions of $\exp(\lambda_k t)$ on both sides of the equalities, we obtain the algebraic system:

$$\lambda_k \beta_i^k = \sum_{j=1}^{n} k_{ji} \beta_j^k \qquad i, k = 1, \ldots, n \qquad (2.15)$$

We have thus all possible relations between the k_{ij} and the unknown β_j^k. They are not inevitably independent. But the n^2 relations (2.15) are of maximum degree 2 in the unknowns. As seen previously, in some simple cases, the equations may be reduced to a linear system but there is no way of generalizing this result to every compartmental model. Nevertheless, a first practical method appears for identifying the k_{ij} and the unknown β_j^k. The numerical algorithm is as follows:

a) First determine (by techniques shown previously) the characteristic values λ_i and the β_j^k associated with the measured $x_i(t)$, $(i = 1, \ldots, p \leqslant n)$.

b) Choose an arbitrary set of $k_{ij} \geqslant 0$ and introduce the functional:

$$J = \sum_{i,j=1}^{n} [\lambda_k \beta_i^k - \sum_{j=1}^{n} k_{ji} \beta_j^k]^2$$

Minimizing J as a function of the β_j^k leads to:

$$\partial J / \partial \beta_j^k = 0$$

for all j,k corresponding to unknown variables β_j^k. This last algebraic system is <u>linear</u> and includes as many equations as unknowns. The resolution gives an optimal set $(\beta_j^k)^*$.

c) Then we put $\beta_j^k = (\beta_j^k)^*$ in the functional J and we minimize J as a function of the k_{ij}. The solution is obtained by solving the <u>linear</u> algebraic system:

$$\partial J / \partial k_{ji} = 0$$

We obtain an optimal solution denoted $(k_{ij})^*$

d) Now set $k_{ij} = (k_{ij})^*$ in functional J and return to step b) and minimize J as a function of the β_j^k.

The algorithm is continued until we obtain a value of J less than or equal to a fixed $\epsilon > 0$. We can also decide to stop the algorithm when two successive values for $(k_{ij})_n^*$ and $(k_{ij})_{n+1}^*$ and for $(\beta_j^k)_n^*$ and $(\beta_j^k)_{n+1}^*$ are very close.

This method was tried successfully on concrete problems from pharmacology [13]. The problems were related to the identification of models (2, 3 or 4 compartments) associated with drugs acting on the heart (blood pressure and heart rate). Convergence is practically always assured. Theoretically it is not difficult to prove the convergence of $(k_{ij})^*$ and $(\beta_j^k)^*$ because J is a positive and decreasing function. J is also a strictly convex function of the β_j^k or the k_{ij} (separately!). The convergence towards an acceptable solution will be assured if $J \to 0$. The last condition is easy to verify on a calculator. The method can prove the existence of a solution but it is ineffective when studying its uniqueness. For uniqueness, the previous techniques (based on Laplace transforms or the analytical study of algebraic relations) are more successful.

Success cannot be guaranteed since there are no general theorems concerning the solution of non-linear algebraic systems [2]. Consider, for example, a variant of the model described in Figure 4.

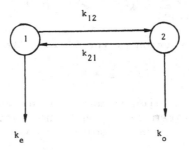

FIG. 6 .

The equations are:

$$\dot{x}_1 = - (k_{12} + k_e)x_2 + k_{21}x_2$$

$$\dot{x}_2 = k_{12}x_1 - (k_{21} + k_0)x_2 \qquad (2.16)$$

$$x_1(0) = \alpha \text{ fixed,} \qquad x_2(0) = 0$$

Is it possible to identify the four k_{ij} from the experimental data $x_1(t)$?

In this case, it is easy using the Laplace transform. We obtain:

$$s\hat{x}_1 - \alpha = - (k_{12} + k_e)\hat{x}_1 + k_{21}\hat{x}_2$$

$$s\hat{x}_2 = k_{12}\hat{x}_1 - (k_{21} + k_0)\hat{x}_2$$

Eliminating $\hat{x}_2$ by substituting:

$$\hat{x}_2 = k_{12}\hat{x}_1/(s + k_0 + k_{21})$$

gives:

$$(s + k_{12} + k_e)\hat{x}_1 + k_{12}k_{21}\hat{x}_1/(s + k_0 + k_{21}) = \alpha$$

Then:

$$x_1/\alpha = (s + k_0 + k_{21})/((s + k_{12} + k_e)(s + k_0 + k_{21}) + k_{12}k_{21})$$

involves only three relations for the k_{ij}. That is to say:

$$k_0 + k_{21} = A$$

$$k_{12} + k_e + k_0 + k_{21} = B$$

$$(k_{12} + k_e)(k_0 + k_{21}) + k_{12}k_{21} = C$$

It is not sufficient to find a solution. Even a very simple compartmental model may not be unique. When a unique solution is not directly found from the data, it must be established by using some other method (addition of some criterion).

Let us now consider the compartmental model in

Figure 5, but with $C_1(t)$ given. C_1, the concentration, is related to the quantity in the compartment by:

$$C_1 = x_1/V_1$$

where V_1 (volume of the compartment) is assumed constant.

By rewriting the differential system with new unknowns:

$$C_i = x_i/V_1$$

it is possible to use one of the techniques already described to show relations between the unknowns. It is easy to show that if k_a can be chosen a unique solution is obtained. However, the previous relation:

$$\dot{x}_1(0) = k_a D \quad \text{becomes} \quad \dot{C}_1(0) = (k_a/V_1)D$$

and therefore $C_1(0)$ and D only determine k_a/V_1. The value of k_a is undetermined and may be equal to $-\lambda_1$, $-\lambda_2$, $-\lambda_3$. The other methods, and in particular those giving the algebraic system of the second degree, confirm this result. Three solutions are possible. Sometimes it is possible to eliminate one or two of the possibilities because they lead to $k_{ij} < 0$, which is impossible from the biological point of view. However, when there is a multiplicity of solutions, a choice criterion is necessary.

2.2 The uniqueness problem

For a biologist it may be impossible to choose between an infinite, or even a finite, number of compartmental models. What can a mathematician propose in such a situation? As seen previously non-uniqueness results from insufficient observations. Sometimes it is possible to obtain some other data, but it is often impossible to do new experiments. Then two steps are possible.

(i) Use a system model which is drastically reduced so that simplified equations for which the solutions will be unique can be obtained.

(ii) Introduce some new relations to ensure uniqueness. For this the introduction of an optimal criterion is very useful.

The first technique requires each situation to be considered separately; there are no general rules to simplify a mathematical model. Simplifications depend on the biological properties and on approximations. Each problem needs particular attention. The second idea may be formalized more easily. By analogy with the working of some physiological systems (muscle, heart, . . .), a possible hypothesis exists in asserting that the compartmental models expend minimum energy [11], [7], [30]. Now for each compartment, the energy is proportional to the quantity of the substance it contains. For the whole system, the criterion to choose, may be:

$$J = \sum_{j=1}^{n} \int_{0}^{\infty} x_j^2(t)dt \qquad (2.17)$$

for an n-compartment system. Of course, one may add some coefficients $\alpha_i > 0$ to the sum, if necessary. In (2.17) we have $\alpha_i = 1$. Proof of the validity of (2.17) is not easy because we need over defined systems. For example, let us consider the system shown in Figure 6 where we know that uniqueness is not assured with the knowledge that we have of $x_1(t)$. Suppose that $x_2(t)$ is also measured. Then it is easy to show the uniqueness of identification using the previous methods. Analytical calculus of the set (k_{ij}) is therefore possible. From now on we only consider the data $x_1(t)$ and we determine the (k_{ij}) by minimizing the functional J. The optimal set obtained can be compared to that calculated with the given x_1 and x_2 and the validity of the criterion may be verified.

Unfortunately this situation does not exist, or at least could not be found, in the literature. Therefore the proof of the criterion in (2.17) remains an open problem for mathematicians. However, there are two small numerical results which go towards a verification of the criterion.

(i) Firstly, a practical example from the classical literature was treated using the criterion of (2.17). The 3-compartment model, as shown in Figure 5, is relevant to the drug being studied, where $x_0(0) = D$ is known and $C_1(t)$, the concentration in compartment 1, is measured. In this case uniqueness is not assured and three solutions are possible corresponding to $k_a = -\lambda_i$ (i = 1, 2, 3). The author, R. Wagner [72], chose the intermediate value for k_a.

The criterion (2.17) was used to choose from the three possible solutions and found the same one.

(ii) The same compartmental system (Figure 5) was used in another practical example of a study of a drug acting on the heart. $C_1(t)$ was given and we calculated the k_{ij} by minimising criterion (2.17), and the greatest value for k_a was obtained.   In this case biological considerations confirmed the choice.   In most cases of drug action it is the greatest possible value of k_a which must be chosen.

These two examples give a partial proof of the usefulness of the criterion.   However, the matter remains partially open.

What about uniqueness when using criterion (2.17)?   It can be seen that J is a strictly convex function of the unknown t_{ij} coming from the matrix $T = (t_{ij})$.   In fact, $x_i(t)$ in (2.17) are replaced using the relation $x = Ty$, where y is the vector of components $\exp(\lambda_k t)$.   Also, J is positive and satisfies:

$$\lim_{t_{ij}^2 = \infty} J = \infty$$

Classical results [20] show that J must have one unique optimum (t_{ij}^*).   In general some constraints have to be added corresponding to the relations $BT = \Lambda_1,$ $T.1 = \alpha.$ This requires the minimization of J on a convex set defined by the linear relations on t_{ij}.

The more complicated compartmental system of Figure 7 was identified using the criterion J.

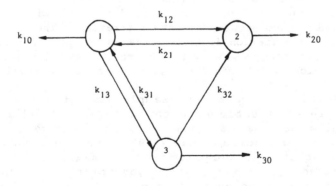

FIG. 7 .

Where $x_1(t)$ and $(x_2 + x_3)(t)$ are measured with $x_1(0) = \alpha$, $x_2(0) = x_3(0) = 0$ it is not difficult to show non-uniqueness. However, minimising J by a numerical optimization technique gives a unique solution.

2.3 Numerical methods for identification

For two or three compartments the k_{ij} can be explicitly calculated (when uniqueness is assured) from the algebraic relations obtained using one of the methods described in Section 2.2. Classical works [72] give explicit formulae where the k_{ij} are expressed as functions of the initial conditions and of the β_i^k obtained from the experimental data:

$$x_i(t) = \sum_{i=1}^{n} \beta_i^k \exp(\lambda_k t)$$

The interested reader will be able to obtain these expressions for himself. They are adequate for solving the simple systems described in the foregoing. When the number of compartments is greater than three, numerical techniques are necessary, involving optimization techniques that will be developed in the next section. Consider a general compartmental system:

$$\dot{x}_i = \sum_{j=1}^{n} k_{ji} x_j \qquad (2.18)$$

$$x_i(0) = \alpha_i \quad \text{(fixed)}$$

with observations (experimental data) given by:

$$z = B \cdot x, \quad B \text{ a known matrix} \qquad (2.19)$$

The first numerical method depends on the numerical resolution of system (2.18). The algorithm is as follows:

(a) First fix a set of k_{ij} and solve the differential system using a numerical method (see Remark below).

(b) Then introduce the functional:

$$J = \sum_{i=1}^{p} \int_{0}^{\infty} (z_i^c - z_i)^2 \, dt \qquad (2.20)$$

using the p functions $z_i(t) \not\equiv 0$. In (2.20), z_i^c designates the function $z_i(t)$ calculated from (2.18) when the k_{ij} are fixed.

(c) The functional (2.20) depends on k_{ij} through the medium of the z_i^c, which are linear combinations of $x_1, \cdots, x_n$. It is therefore possible to minimise J as a function of the k_{ij}.

Min $J = J(k_{ij}^*)$ leads to an acceptable solution at (k_{ij}^*) if we also have $J(k_{ij}^*) = 0$. In practice it is sufficient to satisfy $J(k_{ij}^*) \leqslant \epsilon$, where ϵ is a small parameter depending on the precision of measurement.

Obtaining (k_{ij}^*) satisfying $J(k_{ij}^*) \leqslant \epsilon$ proves the existence of a numerical solution to the identification problem. Proof of uniqueness can only come from a theoretical study.

Remark All numerical methods for solving differential systems are based on very simple ideas [24], [38]. Consider the general system:

$$\dot{x}_1 = f_1(x_1, x_2, t)$$

$$\dot{x}_2 = f_2(x_1, x_2, t), \quad t \geqslant 0 \qquad (2.21)$$

$$x_1(0) = \alpha_1, \quad x_2(0) = \alpha_2$$

A first approach is to "discretise" the system (2.21) by approximating the derivatives $\dot{x}_1$ and $\dot{x}_2$ by:

$$(x_i(t + h) - x_i(t))/h$$

We obtain the new functional system:

$$(x_1(t+h) - x_1(t))/h = f_1(x_1(t), x_2(t), t)$$
$$(2.22)$$
$$(x_2(t+h) - x_2(t))/h = f_2(x_1(t), x_2(t), t)$$

which can be transformed in an algebraic system in which t = nh, where h is, of course, small (tending to zero) and n is a positive integer. We obtain:

$$x_1((n + 1)h) = x_1(nh) + hf_1(x_1(nh), x_2(nh), nh)$$

$$x_2((n + 1)h) = x_2(nh) + hf_2(x_1(nh), x_2(nh), nh) \quad (2.23)$$

$$x_1(0) = \alpha_1, \quad x_2(0) = \alpha_2$$

Equations (2.23) make it possible to calculate successively: $x_1(h)$, $x_2(h)$, $\cdots$ $x_1(ph)$, $x_2(ph)$, $\cdots$, making n = 0, 1, $\cdots$,p-1, $\cdots$

   More sophisticated methods [24] (Runge-Kutta and others) differ only in making better approximations to the derivatives.   Well-known books are available on this important subject [24], [38].
   Another approach, very close to the previous one and from a certain point of view more general, is to consider a regular (class c^1) approximation for x_1 and x_2. That is to say:

$$x_i = \sum_{p=1} c_p^i \, a_p(t) \quad (2.24)$$

where the $a_p(t)$ are known functions giving an approximation to x_1 and x_2. They may be polynomial functions, exponentials [25], or spline functions [49]. Substituting (2.24) in (2.21) gives:

$$\sum_{p=1}^{n} c_p^i \, \dot{a}_p(t) = f_i(\sum c_p^1 a_p(t), \sum c_p^2 a_p(t), t), \quad i = 1, 2$$

$$\sum_{p=1}^{n} a_p(0)c_p^i = \alpha_i$$

The unknowns are now the c_p^i. They can be determined with an optimization technique using the functional:

$$J_1 = \sum_{i=1}^{2} [\int_0^\infty (\sum_{p=1}^{n} c_p^i \dot{a}_p - f_i(\quad))^2 dt] + \sum_{i=1}^{2} (\sum a_p(0)c_p^i - \alpha_i)^2$$

J_1 is minimised as a function of the coefficients c_p^i.

A variant of the numerical identification method is as follows. Identification of the $n(n - 1)$ coefficients k_{ij} can be reduced to the identification of a single parameter. For example, if we have the set of three parameters k_1, k_2 and k_3 we may write:

$$k_1 = a\theta_1\cos\theta_1, \qquad k_2 = a\theta_1\sin\theta_1$$

$$\theta_1 = a\theta\cos\theta, \qquad k_3 = a\theta\sin\theta, \qquad \theta \geq 0$$

where a is a parameter going towards zero.

The method depends on the properties of the Archimedean spiral and can be generalised to n variables. Its working will be seen later in connection with global optimization [11]. It reduces the identification of k_1, k_2, k_3 to the single variable θ. Making the change of variables in J as defined in (2.20) gives a new optimization problem with the variable θ. It is an important simplification allowing faster determination of the optimum θ and hence the k_i values.

Another possibility for identification comes from the last approximation of the earlier remark. We obtain a functional system as below:

$$F_i(c_p^i, k_{ij}) = 0 \qquad i = 1, \ldots, n$$

To solve this for the k_{ij} and the unknown c_p^i we can introduce yet another functional:

$$J_2 = \sum_{i=1}^{n} F_i^2$$

and minimise it as a function of the k_{ij} and the unknown c_p^i. As before, there ought to be a variant in which the unknowns are related through a change of variables. Then a minimization problem with a single variable is obtained and can be solved by a classical method.

A final remark is that we can identify the system represented by (2.18) without solving the differential system. We introduce the functional:

$$J_3 = \sum_{i=1}^{n} \int_0^\infty \alpha_i(\dot{x}_i - \sum k_{ji}x_j)^2 dt + \sum_{i=1}^{n} \beta_i \int_0^\infty (z_i - (Bx)_i)^2 dt \qquad (2.25)$$

where $(Bx)_i$ is the i-th component of the vector z of observations. The α_i and β_i in (2.25) are weighting parameters that may be chosen by the mathematician to facilitate the minimization of J_3. The minimization is as a function of the k_{ij} and the unknown coefficients C_p^i if we substitute the approximations (2.24) for the functions x_i in J_3. It can be seen that J_3 is a second degree polynomial in the k_{ij}, and, separately, in the C_p^i. Thus the algorithm described in Section 2.1 can be applied. At each step we write $\partial J_3/\partial k_{ij} = 0$ or $\partial J_3/\partial C_p^i = 0$, so that the result is obtained by solving a succession of linear equations.

2.4 About the non-linear case

When non-linear models have to be considered there are extra difficulties. Theoretical results obtained in earlier sections cannot be generalised. This is particularly true of those pertaining to existence and uniqueness. For example, we cannot write $x_i(t)$ as a linear combination of exponentials. The property is only approximately true. Thus we can approximate $x_i(t)$ but it is not possible to equate the coefficients of $\exp(\lambda_k t)$ on the two sides of equations because the non-linearities modify the set $(\exp(\lambda_k t))$. All numerical methods described previously are easy to adapt to the non-linear case. Hence they can prove the existence of identification. Only special treatments adapted to each concrete case can be used to demonstrate uniqueness. In some cases uniqueness is studied considering linear algebraic systems.

Consider the following system:

$$x = a_1 x + b_1 y + a_2 x^2 + b_2 y^2$$
$$y = c_1 x + d_1 y + c_2 x^2 + d_2 y^2$$

(2.26)

These equations are useful in modelling many biological systems, especially where hormones play a part or where there are agonist and antagonist regulators [21], [74].

Identification of a_i, b_i, c_i, d_i (i = 1, 2) is equivalent to the resolution of an algebraic linear system if we know $x_1(t)$ and $x_2(t)$ from observations. In fact, considering the data $x(t_j)$, $y(t_j)$, j = 1, . . ,m and introducing the positive function:

$$J_4 = \sum_{j=1}^{m} [\dot{x}(t_j) - a_1 x(t_j) \ldots - b_2 y^2(t_j)]^2$$

$$+ \sum_{j=1}^{m} [\dot{y}(t_j) - c_1 x(t_j) \ldots - d_2 y^2(t_j)]^2 \qquad (2.27)$$

and introducing some approximations for $\dot{x}(t_j)$ and $\dot{y}(t_j)$, we can minimise J_4 as a function of a_i, b_i, c_i, d_i ($i = 1, 2$). Since J_4 is a second degree polynomial, a neceseary and sufficient condition for optimization is given by:

$$\partial J_4/\partial a_i = \partial J_4/\partial b_i = \partial J_4/\partial c_i = \partial J_4/\partial d_i = 0 \qquad (2.28)$$

Since this is a linear algebraic system we have a solution to the identification problem.

In general, when unknown parameters appear in a linear manner in the mathematical equations, the identification problem can be reduced to the resolution of a linear system by introducing a quadratic functional such as (2.27).

There are no more general results for the non-linear case. Nevertheless, an important remark which will be very useful is the following:

Consider experimental data from a biological system, possibly non-linear, as follows:

$$x_1(t), \ldots, x_p(t)$$

and with initial conditions:

$$x_1(0) = D, \quad x_1(0) = 0 = x_2(0) = \ldots x_n(0).$$

We can choose the smallest n such that $x_1, \ldots, x_p$ may be approximated by a linear combination such as the following:

$$x_i(t) = \sum_{p=1}^{N} c_p^i \exp(\lambda_p t) \qquad (2.29)$$

where the λ_p are the same for all x_i (λ_p independent of i).

It follows that our experimental data are consistent with an n-compartment model whose minimal structure may be identified using previous methods. The non-linearity cannot be identified without supplementary experimental data. Suppose, further, that we know experimental data $x_1, \cdot \cdot \cdot, x_p(t)$ associated with doses D_i, $i = 1, \cdot \cdot \cdot, m$, of various magnitudes. We can set up the system (2.29) for each D_i, and thus express the approximations to the C_p^i and the λ_p as functions of D and write:

$$x_i(t) = \sum_{p=1}^{N} C_p^i (D) \exp(\lambda_p(D).t) \qquad (2.30)$$

This allows the biological system to be considered as a <u>non-linear</u> compartmental system. The non-linearities are introduced as functions of D. Hence the k_{ij} of the model will depend on D. Later, we shall see that this approximation is almost completely general as a means of representing non-linearities arising in compartmental systems even when the input u(t) is more general than the Dirac delta-function. In fact a general function u(t) can be approximated by step functions and these can be approximated by Dirac functions.

In conclusion it should be noted that non-linear systems are approximated by successive linear systems depending on the dose D. For D fixed, the associated compartmental model is linear.

Now it is necessary to examine optimization methods whose usefulness in biomathematics stems from the foregoing.

2.5 Optimization techniques

2.5.1 GENERAL CONSIDERATIONS

In the earlier sections many identification problems depended on the minimization of functionals. In general, when we have N experimental points (x_i, y_i), $i = 1$, $2, \cdot \cdot \cdot, N$ it can be useful to find a function such that the graphic representation goes through the experimental data (x_i, y_i). One can look for a regular function F defined by some parameters. For instance, F may be a polynomial function:

$$F(x) = \sum_{i=0}^{m} a_i x^i$$

an exponential function:

$$F(x) = \sum_{k=1}^{m} a_k \exp(\lambda_k x), \text{ and so on.}$$

The unknown parameters a_i, λ_i have to be identified as well as possible. This means that the following criterion J:

$$J(a_1, \ldots, a_p) = \sum_{i=1}^{N} (y_i - F(a_1, \ldots, a_p, x_i))^2 \qquad (2.31)$$

must be minimised as a function of $a_1, \ldots, a_p$; that is to say:

$$\underset{a_i \in R}{\text{Min}} \quad J = J(a_1^*, \ldots, a_p^*)$$

The structure of F is always specified that only parameters (real or complex) have to be identified. An optimization problem may be reduced to the following general formalism:

$$\underset{x \in K}{\text{Min}} \quad J(x)$$

where $K = [x \in R^n \mid g_i(x) = 0, i = 1, \ldots, m$ and
$$k_j(x) \geqslant 0, j = 1, \ldots, r]$$

The functions k_i and g_i defining the set K are called constraints. If $K = R^n$ the optimization problem is said to be without constraints. It is easily seen that a maximization problem Max G(x) is equivalent to a minimization problem by setting $J = -G$, and we have Max $G(x) = -$ Min $J(x)$. It is therefore only necessary to study minimization techniques.

For considering existence and uniqueness of optimal solutions it is useful to introduce partial derivatives. Let us consider J to be twice differentiable on R^2. Suppose that (x^*, y^*) is a strict local minimum; that is to say:

There is a neighbourhod $V(x^*, y^*)$ such that $(x, y) \in V$ implies $J(x^*, y^*) < J(x, y)$. The expansion of J by Taylor's theorem in the neighbourhood of (x^*, y^*) is given by:

$J(x*+h, y*+k) =$

$J(x*, y*) + J'(x*, y*)\binom{h}{k} + (1/2)(h,k).J"(x*, y*)\binom{h}{k} + O(h^3, k^3)$

where $J'(x*, y*) = (\partial J/\partial x(x*, y*), \partial J/\partial y(x*, y*))$

and $J"(x*, y*) = H(x*, y*) =$

$$\begin{bmatrix} \partial^2 J/\partial x^2(x*, y*) & \partial^2 J/\partial x \partial y(x*, y*) \\ \partial^2 J/\partial y \partial x(x*, y*) & \partial^2 J/\partial y^2(x*, y*) \end{bmatrix}_{(2.32)}$$

is the Hessian of J at the point (x*, y*).

For h and k sufficiently small we have, taking into account the property of (x*, y*):

$J'(x*, y*)\binom{h}{k} + (1/2)(h,k).J"(x*, y*)\binom{h}{k} + O(h^3, k^3) > 0$

(2.33)

Since h and k may take positive or negative values the strict positivity of (2.33) is only assured when:

$J'(x*, y*) = 0$

If this condition holds, (x*, y*) is a strict local minimum if the following is satisfied:

For every h, k $(1/2)(h,k).J"(x*, y*)\binom{h}{k} > 0$ (2.34)

which means that the Hessian is a positive definite matrix [3a]. If the matrix is only positive semi-definite the minimum is not strict. The following theorem can therefore be stated:

Theorem 2.5.1 Since J is a twice differentiable function in an open set U of R^n we can say:

(i) If J has a local minimum at x* then $J'(x*) = 0$ and the derivative $J"(x*)$ is positive semi-definite, that is to say:

For all $w \in U$, $w^T J"(x*)w \geqslant 0$ where w^T is the transpose vector of w.

(ii) If $J'(x^*) = 0$ and J is twice differentiable in U and
if $J''(u)$ is semi-positive definite in U then J has a
relative minimum at x^*.

Now the notion of convexity will be introduced to
complete the results on optimization.

A set K is said to be convex if for each $(x,y) \in K^2$ the
following relation is true:

for $t \in [0,1]$, $x \in K$, $y \in K \Rightarrow (1 - t)x + ty \in K$

J is said to be convex in the convex set K if for $x,y \in K$,
$\theta \in [0,1]$, $x \neq y \Rightarrow J(\theta x + (1 - \theta)y) \leqslant \theta J(x) + (1 - \theta)J(y)$
The previous theorem gives the necessary conditions to
ensure that x^* is a local minimum in U. These results are
only valid for unconstrained optimization problems. When
the constrained problem is considered,

$$\text{Min} J(x) \qquad\qquad (2.35)$$
$$x \in K$$

conforms to the following.

Theorem 2.5.2 Let x^* be a local minimum of J in K and J be
differentiable at x^*. We have:

$$J'(x^*)(y - x^*) > 0 \qquad\qquad (2.36)$$
for all $y \in K$.

Setting $z = x^* + y \in K$ we can write, for $\theta \in [0,1]$:

$$J(x^* + \theta y) = J(x^*) + \theta J'(x^*) + \theta \|y\| \epsilon(\theta y)$$

with $\epsilon(\theta y) \to 0$ when $\theta \to 0$.

From the definition of the minimum, $\theta J'(x^*) \geqslant 0$ and
therefore $J'(x^*)(z - x^*) \geqslant 0$ for all $z \in K$.

A corollary may be proved [46] using the convexity
properties of J.

Corollary 2.5.2 Let J be convex in the convex set K, and
differentiable.

J convex $\iff \forall\ x,y \in K \Longrightarrow J(y) \geqslant J(x) + J'(x)(y - x)$

J strictly convex $\iff \forall\ x,y \in K \Longrightarrow J(y) > J(x) + J'(x)(y - x)$

The strict convexity is satisfied if our inequality is strict.

When J is twice differentiable on K then:

J convex $\iff J''(x)$ positive semi-definite for all $x \in K$.

The following is an important result in optimization theory and practice.

<u>Theorem</u> 2.5.3 Let J be a function from R^n in R and $K \neq \phi$ a closed set of R^n. If one of the next two conditions is satisfied:

(a) K is bounded

(b) $\lim_{\|x\| \to \infty} J(x) = +\infty$

then the problem $\underset{x \in K}{\text{Min}}\ J(x)$ has at least one solution x^*:

$$(\underset{x \in K}{\text{Min}}\ J(x) = J(x^*))$$

The proof is easy. If $K \subset R^n$ is bounded, it is also compact and the result follows. In the other case choose $a \in K$, then from (b) there exists r such that for all $x \in R^n$ with $\|x\| \geqslant r$ we have $J(a) < J(x)$. The problem has one solution if and only if we look for a minimum in the set:

$$K \cap [x \in R^n;\ \|x\| < r]$$

This is a bounded set and we return to the first case.

This theorem ensures existence of a solution. For uniqueness it is possible to use the following result.

<u>Theorem</u> 2.5.4 Where K is a convex set in R^n and $J:R^n \to R$ a convex function, every local minimum is also a global one. Furthermore, if J is strictly convex, it has at most one minimum which is a strict minimum.

The proof is easy. Let u be a local minimum of J, and $v \in K$. J is convex and therefore, by definition:

$$J(\theta v + (1 - \theta)u) \leqslant \theta J(v) + (1 - \theta)J(u)$$

This inequality implies:

$$J(\theta v + (1 - \theta)u) - J(u) \leqslant \theta(J(v) - J(u))$$

and for small θ such that:

$$J(\theta v + (1 - \theta)u) - J(u) \geqslant 0$$

(the last possible because u is a relative minimum), we have:

$$\theta(J(v) - J(u)) \geqslant 0, \quad \text{that is} \quad J(v) \geqslant J(u).$$

The second part can be demonstrated using the same technique and noting that for a strictly convex function a local minimum is necessarily strict.

2.5.2 NUMERICAL METHODS [46]

It is rarely possible to calculate the optimum by analytical methods. When an analytical technique is used we have the following relations:

$$\partial J/\partial x_i = 0 \quad , \quad i = 1, \ldots, n, \quad \text{where } v = (x_i)$$

The system is easy to solve when it is linear. Otherwise numerical methods have to be used. A technique for a single variable is described first. The problem is to find:

$$x^* \in [a_0, b_0]$$

such that:

$$f(x^*) = \underset{x \in [a_0, b_0]}{\text{Min}} f(x) \tag{2.37}$$

The dichotomic technique decreases the length interval containing the minimum in successive steps so long as the length is greater than a fixed threshold ϵ.

Starting with the interval $[a_n, b_n]$ we consider the point $c_n = (a_n + b_n)/2$ in the middle of $[a_n, b_n]$ and the points x_1^n, x_2^n defined by $x_1^n = c_n - \epsilon/2$, $x_2^n = c_n + \epsilon/2$

$$\text{If } f(x_2^n) < f(x_1^n) \text{ then } a_{n+1} = x_1^n, \ b_{n+1} = b_n$$

$$\text{If } f(x_2^n) > f(x_1^n) \text{ then } a_{n+1} = a_n, \ b_{n+1} = x_2^n$$

In each step the length interval is approximately divided by 2. This method requires only continuity whereas Newton's method requires differentiability of the function.   It however can be used with several variables.   The principle is to calculate numerically a solution of:

$$\partial J/\partial x_i = 0 \ , \ i = 1, \ . \ . \ ,n$$

which is a non-linear system of algebraic equations.
        With one variable the determination of x such that f(x) = 0 goes through the following sequence:

> (x_i) is defined by the recurrent relation:
> x_0 arbitrary
> $x_{i+1} = x_i - f(x_i)/f'(x_i)$

In other words x_{i+1} is the intersection of the tangent at x_i with the horizontal axis.   The technique may be generalized to p equations with p unknowns:

$$f_i(x) = 0, \quad i = 1, \ . \ . \ ,p; \quad x \in R^p \qquad (2.38)$$

by introducing the Jacobian matrix:

$$Df(x) \quad = \quad \begin{bmatrix} \partial f_1/\partial x_1 & \cdots & \partial f_p/\partial x_1 \\ \cdot & & \cdot \\ \cdot & & \cdot \\ \cdot & & \cdot \\ \partial f_1/\partial x_p & \cdots & \partial f_p/\partial x_p \end{bmatrix}$$

The sequence $(x_i) \in R^p$ is therefore defined by:

$$x_{i+1} = x_i - \lambda_i [Df(x_i)]^{-1} f(x_i)$$

where the sequence $(\lambda_i) \in R$ determines the steps.   For example one may choose the optimal λ_i for which the function J is minimal along the direction:

$$x_i - \lambda [Df(x_i)]^{-1} f(x_i)$$

This is equivalent to solving at each iteration a unidimensional problem, as follows:

$$J(x_i - \lambda^* [Df(x_i)]^{-1} f(x_i)) \leqslant J(x_i - \lambda [Df(x_i)]^{-1} f(x_i))$$

for all $\lambda \in R$. The dichotomic technique can be used to find λ^*. The "descent" techniques are more practical.

2.5.3 DESCENT METHODS [20]

The gradient technique belongs to this important class which depends on constructing a sequence (x_i) satisfying:

$$J(x_{k+1}) < J(x_k) , k = 1, .2, \cdots$$

We hope that x_k converges on the desired optimum x^*. This occurs under some conditions. In the gradient method we want a maximal difference $J(x_{k+1}) - J(x_k)$. We know that the gradient vector $(J'(x_{k+1}))$ maximises this difference and therefore at each step we want to solve the following unidimensional problem. Find $\lambda \in R$ such that:

$$J(x_k - \lambda J'(x_k)) \leq J(x_k - \lambda_k J'(x_k)) \qquad (2.39)$$

for all λ_k.

Many variants of the gradient method [18]; [46] are possible (for instance the conjugate gradient technique) and we shall now study a discrete form of the technique [12], [18]. The problem is to solve numerically the optimization problem:

$$\underset{x \in R^n}{Min} \; f(x) \; = \; \underset{x_1, \cdots, x_n}{Min} \; f(x_1, \cdots, x_n) \qquad (2.40)$$

We shall develop a direct method whose algorithm does not use the derivatives. For simplicity we shall use $n = 2$, but the method is general.

First an arbitrary point $(x_0, y_0) \in R^2$ and a step h are chosen. Then among the three points (x_0, y_0), $(x_0 \pm h, y_0)$ we retain the one which gives the minimal value of f. Starting with this point (say (x_0+h, y_0)) we choose among $(x_0 + h, y_0)$, $(x_0 + h, y_0 \pm h)$ the point giving the smallest value of f. Then we return to the first variable x and so on until a stationary point (x_h, y_h) is obtained.

It satisfies:

$$f(x_h \pm h, y_h) \geq f(x_h, y_h)$$
$$\qquad (2.41)$$
$$f(x_h, y_h \pm h) \geq f(x_h, y_h)$$

From this stationary point the process is iterated with a smaller value of h (in practice usually h/2). The iterative algorithm is stopped when the results of two successive iterations are close. Convergence depends on some hypotheses.

Theorem 2.5.5   If f is strictly convex and continuously differentiable and satisfies:

$$\lim_{x^2+y^2 \to \infty} f(x,y) = +\infty$$

then f has a unique optimum (minimum) and furthermore the method converges when $h \to 0$. In other words $x_h \to x^\dagger$, $y_h \to y^\dagger$ (when $h \to 0$) with $f(x^\dagger,y^\dagger) = \text{Min } f(x,y)$.

For the proof, note that the algorithm does not require derivatives to exist but the continuity of derivatives is necessary to prove convergence. There are, in fact, counter examples of non-convergence when f is not continuously differentiable. The first part of our theorem is obvious and results from previous considerations. Then the sequence (x_h,y_h) is bounded because if it were not we would have $x_h^2 + y_h^2 \to \infty$ for some subsequence and therefore $f(x_h,y_h) \to \infty$. This last is in contradiction with $f(x_h,y_h) \leqslant f(x_0,y_0)$ finite. A convergent subsequence can be extracted from the sequence (x_h,y_h):

$$x_{h'} \to x^*$$

$$y_{h'} \to y^* \qquad \text{when } h' \to 0$$

We must show that $x^* = x^\dagger$ and $y^* = y^\dagger$.

Using the finite increments theorem and the definition of the stationary point leads to:

$$h'f_x'(x_{h'} + \Theta h', y_{h'}) \geqslant 0 \qquad\qquad \text{with } \Theta \in [0,1]$$

$$- h'f_x'(x_{h'} - \Theta_1 h', y_{h'}) \geqslant 0 \qquad\qquad \Theta_1 \in [0,1]$$

$$h'f_y'(x_{h'}, y_{h'} + \Theta'h') \geqslant 0 \qquad\qquad\qquad\qquad (2.42)$$

$$- h'f_y'(x_{h'}, y_{h'} - \Theta_1'h') \geqslant 0 \qquad\qquad \Theta', \Theta_1' \in [0,1]$$

Dividing (2.42) by h' > 0 and expressing the limits when
h → 0 (h' → 0) give:

$$f'_x(x^*,y^*) \overset{>}{\underset{<}{}} 0 \implies f'_x(x^*,y^*) = 0$$

and $\qquad f'_y(x^*,y^*) \overset{>}{\underset{<}{}} 0 \implies f'_y(x^*,y^*) = 0$

(x^*,y^*) is a minimum of f and because the minimum is unique
we have $x^* = x^\dagger$, $y^* = y^\dagger$. A subsequence converges towards
the global minimum of f. The sequence itself must also
converge on $(x^\dagger,y^\dagger)$.

In Biomathematics the functionals to be minimised are
usually non-convex, and so a numerical optimisation method
obtains a local minimum. Consequently it would be useful to
have a global optimization technique converging on the
global optimum (minimum or maximum). The next paragraph
develops a new and fruitful idea [11], [14].

2.5.4. A GLOBAL OPTIMIZATION TECHNIQUE [11], [14]

Very simple methods exist for looking for the global optimum
of n-variable functions. Unfortunately they cannot be used
on calculators because they need too much machine time.
Some ingenuity is necessary to discover some new method. In
a sense, reducing an n-variable function to a single
variable function seems to be a good approach because very
good optimization techniques for a real variable can be
found in the literature. (See above).

Consider the real function $f(x_1, \ldots ,x_n)$ whose minimum
is wanted. We shall use a transformation based on the
properties of the Archimedean spiral $r = a\theta$ (in polar co-
ordinates). In fact when $a \to 0$, the Archimedean spiral can
be considered as an approximation of Peano's curve and
therefore as an approximation of R^2. Let us describe the
transformation with n = 4. Setting:

$$x_1 = r_1\cos\theta_1 , \quad x_2 = r_1\sin\theta_1 , \quad r_1 = a_1\theta_1$$
$$x_3 = r_2\cos\theta_2 , \quad x_4 = r_2\sin\theta_2 , \quad r_2 = a_2\theta_2$$

$$(2.43)$$

The trick lies in relating (r_1,θ_1) by means of the
Archimedean spiral. Substituting in the function f gives:

$$f(x_1, x_2, x_3, x_4) =$$

$$f(a_1\theta_1\cos\theta_1,\ a_1\theta_1\sin\theta_1,\ a_2\theta_2\cos\theta_2,\ a_2\theta_2\sin\theta_2) = g(\theta_1, \theta_2)$$

The first function f depended on four variables, while the new function depends only on two. A final change of variables is necessary:

$$\theta_1 = r\cos\theta,\quad \theta_2 = r\sin\theta,\quad r = a\theta \qquad (2.44)$$

We obtain a new function $G(\theta)$ defined by:

$$f(x_1, x_2, x_3, x_4)$$

$$= f(a_1 a\theta\cos\theta\cos(a\theta\cos\theta),\ a_1 a\theta\cos\theta\sin(a\theta\sin\theta),$$

$$a_2 a\theta\sin\theta\cos(a\theta\sin\theta),\ a_2 a\theta\sin\theta\sin(a\theta\sin\theta)) = G(\theta)$$

Now it suffices to find the absolute minimum of the function G depending on the single variable θ.

As soon as the optimum θ^* is obtained it is possible to return to the original variables x_1^*, x_2^*, x_3^*, x_4^* by using the transformation formulae (2.43) and (2.44).

Of course x_1^*, x_2^*, x_3^*, x_4^* are close to the true global optimum $x_1^\dagger$, $x_2^\dagger$, $x_3^\dagger$, $x_4^\dagger$ satisfying:

$$\text{Min } f(x_1,\ x_2,\ x_3,\ x_4) = f(x_1^\dagger,\ x_2^\dagger,\ x_3^\dagger,\ x_4^\dagger)$$

Indeed for each $(x_1,\ x_2,\ x_3,\ x_4) \in R^4$ there exists θ_1 and a_1 such that $x_1^{(0)} = a_1\theta_1\cos\theta_1$, $x_2^{(0)} = a_1\theta_1\sin\theta_1$ are close to x_1, x_2. Idem for $x_3^{(0)} = a_2\theta_2\cos\theta_2$, $x_4^{(0)} = a_2\theta_2\sin\theta_2$, which are close to x_3, x_4 for some a_2, θ_2. Then the continuity of the second change of variables $\theta_1^{(0)} = a\theta\cos\theta$, $\theta_2^{(0)} = a\theta\sin\theta$ allow θ and a to be chosen such that $\theta_1^{(0)}$ and $\theta_2^{(0)}$ are close to the previous θ_1 and θ_2 and hence that $(x_1^{(0)},\ x_2^{(0)},\ x_3^{(0)},\ x_4^{(0)})$ is close to $(x_1,\ x_2,\ x_3,\ x_4)$.

Except in the case of n = 2 it is not possible to calculate the maximum distance between a point of R^n and a point associated with our Archimedian transformation. Nevertheless a numerical program (see Appendix 1) has been written to evaluate this maximal distance by a probabilistic technique. It gives an idea of the maximum error associated with the coefficients a_i and demonstrates the "density" of our Archimedean transformation in R^n.

Possible applications include identifying a linear exponential combination such as the following:

$$x(t) = \sum_{i=1}^{n} a_i \exp(-\lambda_i t) \quad , \qquad t \geqslant 0, \; \lambda_i \geqslant 0 \quad (2.45)$$

The positive funtion J depending on a_i, λ_i is:

$$J = \sum_{i=1}^{M} (x(t_j) - \sum_{i=1}^{n} a_i \exp(-\lambda_i t_j))^2 \qquad (2.46)$$

and is minimized as a function of these.

The Archimedean transformation (called Alienor [15]) is used after setting:

$$r_1 = \exp(-\lambda_1), \; r_2 = \exp(-\lambda_2), \; r_3 = \exp(-\lambda_3), \cdot \cdot$$

$$0 < r_1 < 1$$

and

$$r_1 = (1 + th\lambda_1)/2; \quad r_2 = (1 + th\lambda_2)/2; \quad r_3 = (1 + th\lambda_3)/2, \cdot \cdot$$

where th is the hyperbolic tangent.

The mathematical transformation is the following:

$$\lambda_1 = \beta_1 \; \cos\beta_1/c \qquad\qquad a_1 = \beta_1 \; \sin\beta_1/c$$

$$\lambda_2 = \beta_2 \; \cos\beta_2/c \qquad\qquad a_2 = \beta_2 \; \sin\beta_2/c$$

$$\lambda_3 = \beta_3 \; \cos\beta_3/c \qquad\qquad a_3 = \beta_3 \; \sin\beta_3/c$$

$$\cdot \quad \cdot \quad \cdot \quad \cdot \qquad\qquad\qquad \cdot \quad \cdot \quad \cdot \quad \cdot \qquad (2.47)$$

$$\beta_1 = \gamma_1 \; \cos\gamma_1 \; \cos\gamma_2/c \qquad\qquad \beta_2 = \gamma\sin\gamma_1 \; \cos\gamma_2/c$$

$$\beta_3 = \gamma_1 \; \sin\gamma_2/c$$

$$\gamma_1 = \theta \; \cos\theta/c \qquad\qquad\qquad \gamma_2 = \theta \; \sin\theta/c$$

which involve $J = F(\theta)$, $\theta \in R^+$.

The program, written in Basic, is given and discussed in Appendix 1.

For n = 2, a concrete problem was solved using the transformation:

$$\lambda_1 = \text{ABS } (\theta\cos\theta)/100$$

$$\lambda_2 = \text{ABS } (\theta\sin\theta)/100 \tag{2.48}$$

a_1 and a_2 in (2.45) are calculated using least square technique [29]. When n = 3, the following transformation (suggested by empirical attempts) may be used:

$$\lambda_1 = \text{ABS } (\alpha\cos\alpha)/10$$

$$\lambda_2 = \text{ABS } (\alpha\sin\alpha)/10$$

$$\lambda_3 = \text{ABS } (\beta)/10 \tag{2.49}$$

$$\alpha = \theta \cos\theta/10$$

$$\beta = \theta \sin\theta/10$$

The a_i, λ_i are calculated from the experimental data given in Table 2.A. The function x^c (x calculated) is very close to the x found from experiment.

The a_i are included in the formula:

$$x^c(t) = 23.23826 \exp(-0.23768t) + 16.06332 \exp(-8.803965t)$$

We have an error in the coefficients, equal to $\pm$ 3.10%.

Another concrete example, with n = 3, was treated, with values as tabulated in Table 2.B.. This allows the comparison between x^c calculated using the global optimization technique and the x from experiment.

In Table 2.B:

$$x^c(t) =$$
$$-3.42\exp(-1.087t) - 0.535\exp(04.107t) + 3.958\exp(-0.098t)$$

with a global error on coefficients of 2.19%.

t hrs.	x exp.	x^c	t hrs.	x exp.	x^c	t hrs.	x_1 exp.	x_1 calc.
0	36.30	36.30	1	1.05	1.057	1	1.50	1.52
0.08	29.26	27.45	1.5	2.20	2.136	1.5	3.20	3.11
0.17	25.24	25.55	2	2.55	2.62	2	3.50	3.553
0.25	23.34	23.35	3	2.85	2.88	3	3.50	3.56
0.75	19.46	19.65	4	2.70	2.79	4	3.20	3.26
1	18.32	18.00	6	2.15	2.355	6	2.70	2.758
2	14.45	15.45	8	2.00	1.94	8	2.35	2.323
3	11.39	10.75	10	1.55	1.60	10	1.95	1.96
4	8.98	8.90				24	1.20	0.59
5	7.08	7.25	28	0.95	0.27	28	0.90	0.42
6	5.58	6.00	32	0.90	0.19	32	0.55	0.300
			36	0.80	0.13	36	0.40	0.210

| Table 2.A | Table 2.B | Table 2.C |

In Table 2.B it can be seen that the results are good up to t = 10. The last three points correspond to large t and involve small exponentials, and experimental values are not very precise at these points. If necessary, weighting coefficients can be introduced in the functional (2.46) so as to improve some results, especially when t $\geqslant$ 28 hours.

A more complicated problem is to identify a three compartment model, as in Figure 5, (see section 2.1), with initial conditions $x_1(\tau) = x_2(\tau) = 0$ and $x_0(\tau) = D$ where τ is an unknown time lag. This time lag is common when studying drug action such as that of beta-blockers. The model equations are easy to obtain and $x_1(t)$ represents the experimental values. The global optimization technique

(Alienor) gives the a_i and λ_i of the formula:

$$x_1(t) = a_1\exp(-\lambda_1 t) + a_2\exp(-\lambda_2 t) + a_3\exp(-\lambda_3 t) \quad (2.50)$$

We have: $a_1 = -0.687$ $\lambda_1 = 2.7448$
$a_2 = -3.760$ $\lambda_2 = 2.0270$
$a_3 = 4.446$ $\lambda_3 = 0.0858$

Table 2.C allows comparison between x_1 from experiment and x_1 calculated.

The time lag τ is identified using (2.50) and looking for a value of τ such that $x_1(\tau) = 0$. A classical method to resolve $f(x) = 0$ gives the time lag $\tau = 0.7693$ hours (approximately 46 minutes) when $D = 106.39$. Therefore if we choose $k_a = \lambda_3 = 0.0858$ it is possible to determine the exchange parameters k_{ij} analytically. The previous techniques can be applied to obtain:

$$k_{21} = 2.602, \quad k_0 = 2.1385, \quad K_{12} = 0.031556$$

Now we know how to identify a compartmental system. We can briefly discuss its usefulness. We saw that it is important to have a model at ones disposal to simulate the system under various circumstances. Furthermore, the identification of k_{ij} gives information about the biological properties. For example, a large k_{12} going quickly into the action compartment. It is favourable for therapeutics. On the other hand, if k_e (the elimination coefficient) is small, then the substance is slowly eliminated.

A compartmental model can also be useful in treating illness. Optimal possibilities of drug action are easy to see using optimal control theory [3], [65].

Remark 1 A simpler method may be used to solve optimization problems where the function is a non-linear polynomial. Consider a concrete example:

Min $[f(x,y)$
x,y
$$= ax + by + cxy + dx^2y + ex^3y^2 + gx^2y^2 + hx^2y^3 + ix^4y^4]$$

New variables are introduced:

$$u = xy, \quad v = ux \; (= x^2y), \quad w = vu \; (= x^3y^2), \quad z = vy \; (= x^2y^2)$$

$$s = zy \; (= x^2y^3), \quad t = vs \; (= x^4y^4)$$

Then the new optimization problem is considered:

$$\underset{x,y,u,v,\ldots,t}{\text{Min}} \quad J = [ax + by + cu + dv + ew + gz + hs + it$$

$$+ (u - xy)^2 + (v - ux)^2 + \ldots + (t - vs)^2]$$

The numerical algorithm is the following:

Fix (x,y) and minimise the function in terms of $u, v, \ldots, t$. A <u>linear</u> algebraic problem is obtained by writing:

$$\partial J/\partial u = \ldots = \partial J/\partial t = 0$$

giving $u^*, \ldots, t^*$. Fixing $u = u^*, \ldots, t = t^*$ in J, the function J is then minimised in terms of (x,y). A <u>linear</u> system:

$$\partial J/\partial x = \partial J/\partial y = 0$$

has to be solved. It gives the optimal values x^*, y^*.

Going back to the first step where $x = x^*$, $y = y^*$ we obtain a recurrent algorithm giving a decreasing sequence for J. In some cases the convergence can be easily proved.

The main interest of this technique is that at each step a <u>linear</u> problem has to be solved. New variables were introduced to uncouple the non-linearities.

<u>Remark</u> <u>2</u> The Alienor technique can be improved by using an idea (due to A. Guillez of Medimat) which is to put more than two variables on the same Archimedean spiral. Let us consider three variables x, y, z and the following transformation:

$$x = a\theta\cos\theta, \quad y = a\theta\sin\theta, \quad z = a\theta\sin m\theta \qquad (2.51)$$

where m, a real number, has to be chosen. We thus reduce the three variables to a single one, namely θ.

CHAPTER 3

OPTIMAL CONTROL IN COMPARTMENTAL ANALYSIS

3.1 General considerations.

Control variables are needed to act on a compartmental system. In our previous developments $x_1(0) = \alpha$ the injected dose, or the absorbed dose, at time $t = 0$ plays the role of an action variable. For simplicity the mathematical and numerical methods will be developed on a two-compartment system, but the extension to n compartments may be done without difficulty.

Let us consider the model shown in Figure 8:

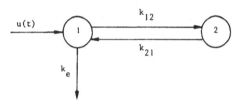

FIG. 8 .

which leads to the mathematical equations:

$$\dot{x}_1 = -(k_{12} + k_e)x_1 + k_{21}x_2 + u(t)$$

$$\dot{x}_2 = k_{12}x_1 - k_{21}x_2$$

$$x_1(0) = x_2(0) = 0$$

(3.1)

where u(t) is the control variable which can be chosen to optimise a fixed criterion. u(t) may be a distribution function [59]. In the previous studies (see Figure 4) a particular u(t) was considered:

$$u(t) = \alpha . \delta_{(0)}$$

where $\delta_{(0)}$ is the Dirac delta-function at $t = 0$.
The following is a possible control problem:

Find u(t) such that (3.1) is satisfied

$$\text{and} \int_0^\infty (x_1(t) - a)^2 \, dt \quad \text{is minimised.} \tag{3.2}$$

In (3.2) a is a fixed constant defined from biological considerations. Minimising the criterion:

$$\int_0^\infty (x_1(t) - a)^2$$

is equivalent to making $x_1(t) \equiv a$ as far as possible. Moreover biological studies have shown that drug action is proportional to the concentration in the blood. Therefore it is logical to introduce a criterion including x_1. If necessary, x_2 may also be included :

$$\int_0^\infty (x_2(t) - A)^2 \, dt$$

(3.2) is a control problem for which different methods (analytical and numerical) exist. They will be presented later. First let us notice an important result. A compartmental system has to be identified before it can be controlled. Nevertheless it is often possible to control a system even if its identification is not unique.

Consider the system in Figure 8 with a supplementary exit k_0 from compartment 2 to the outside. We know that uniqueness cannot be assured (three relations for four unknowns). The Laplace transform gives the new system:

$$s\hat{x}_1 = -(k_e + k_{12})\hat{x}_1 + k_{21}\,\hat{x}_2 + \hat{u}$$

$$s\hat{x}_2 = k_{12}\hat{x}_1 - (k_{21} + k_0)\hat{x}_2$$

Eliminating x_2 using $\hat{x}_2 = k_{12}\hat{x}_1/(s + k_{21} + k_0)$ gives:

$$s\hat{x}_1 = -(k_e + k_{12})\hat{x}_1 + k_{12}k_{21}/(s + k_{21} + k_0)\hat{x}_1 + \hat{u} \quad \text{and thus:}$$

$$\hat{u} = [-k_{12}k_{21}/(s + k_{21} + k_0) + k_e + k_{12} + s]\hat{x}_1 \tag{3.3}$$

We obtain a convolution equation between u and x_1. In fact taking the inverse Laplace transform of (3.3) gives:

$$u = K_* x_1 \quad \text{or} \quad x_1 = K_1 * u$$

where K is the inverse of:

$$[s + k_{12} + k_e - k_{21} k_{12}/(s + k_{21} + k_0)]$$

K_1 is known if we measure $x_1^*(t)$ resulting from an injection $u(t) = \alpha . \delta_{(0)}$. In fact, we have:

$x_1^* = \alpha K_1$ because $\delta * g = g$ for all g. Therefore $K_1 = x_1^*/\alpha$.

For $x_1(t)$ corresponding to an arbitrary u(t) we deduce:

$$x_1 = x_1^*/\alpha * u \qquad (3.4)$$

(3.4) gives a relation between the control u and the measured compartment. This relation is well defined if we have an experimental measure associated with an instantaneous injection. The conclusion is therefore as follows:

The optimal control problem $\quad \underset{u(t)}{\text{Min}} \int_0^\infty (x_1(t) - a)^2 dt \quad$ may be

solved because we know an explicit relation betweem x_1 and u. Uniqueness of model identification is not necessary. The optimal control problem is well set because of the relation (3.4) even if the identification problem does not allow a unique solution. Generalization is possible under some hypotheses.

Let us consider the general system:

$$x(t) = Ax(t) + u(t)$$
$$x(0) = 0 \qquad (3.5)$$

where only the first component of u(t) is different to 0. Again the Laplace transform gives:

$$s\hat{x} = A\hat{x} + \hat{u}.$$

If $(sI-A)^{-1}$ exists then:

$\hat{x}(s) = (sI-A)^{-1}\hat{u}(s)$ and therefore $\hat{x}_1(s) = [(sI-A)^{-1}]_{11}\hat{u}(s)$ where $[(sI-A)^{-1}]_{11} = J_{11}(s)$ represents the term of $(sI-A)^{-1}$ in the first row and first column. In fact $J_{11}(s)$ is a rational fraction because of Cramer's rule: J_{11} is the quotient of two determinants. The numerator is of degree $n - 1$ and the denominator has degree n.
Setting:

$$\hat{x}_1(s) = (p_{n-1}(s,k_{ij})/q_n(s,k_{ij}))\hat{u}(s)$$

or if we prefer:

$$\hat{u}(s) = q_n(s,k_{ij})/(p_{n-1}(s,k_{ij}))\hat{x}_1(s)$$

a decomposition of the fraction q_n/p_{n-1} in simple elements of the form $1/(s + b)$ will allow us to find the original in an explicit manner:

$$u(t) = r_n(t) * x_1(t) \quad \text{where} \quad \hat{r}_n(s) = q_n/p_{n-1} \qquad (3.6)$$

In (3.6) the sign $*$ is the convolution product [59]. If the poles s_k of $\hat{r}_u$ are simple, we obtain the formula:

$$r_n(t) = \sum_{k=1}^{m} (q_n(s_k)/p'_{n-1}(s_k)\exp(s_k t))$$

and (3.6) gives an explicit formula for u(t). As before, it is not necessary to identify each k_{ij} uniquely. In practice, the knowledge of the fraction q_n/p_{n-1} is sufficient.

3.2 A first explicit approach

In this section we are interested in a solution u(t) given by a classical regular function and not by a distribution. As before one can see if it is possible to satisfy $x_1(t) \equiv a$. The condition $x_1(0) = 0$ does not allow $x_1(t) = a$ for all $t \geq 0$, but we may try the function illustrated in Figure 9, where $x_1(t) = a$ for $t \geq h$ and $x_1(t) = (a/h)t$ if $0 \leq t \leq h$.

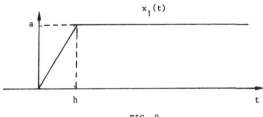

FIG. 9 .

Performing the inverse of the Laplace transform on both sides of (3.3) gives:

$$u(t) = - k_{21}k_{12}\exp(-k_{21}t) \int_0^t \exp(k_{21}\tau)x_1(\tau)d\tau$$
$$+ (k_e + k_{12})x_1(t) + \dot{x}_1(t) \qquad (3.7)$$

Then by integration:

$$u(t) = - k_{12}k_{21}\exp(-k_{21}t) \int_0^t \exp(k_{21}\tau)a/h.\tau d\tau$$

$$+ (k_0 + k_{12})a/h.t + a/h \quad \text{if} \quad 0 \leqslant t \leqslant h$$
$$\qquad (3.8)$$
$$u(t) = ak_e + ak_{12}\exp(-k_{21}t)(\exp(k_{21}h) - 1)hk_{21} \quad \text{if} \quad t \geqslant 1$$

Finally (3.8) is equivalent to:

$$u(t) = ak_{12}/hk_{21}(1 - \exp(-k_{21}t))ak_e t/h + a/h$$
$$\text{if} \quad 0 \leqslant t \leqslant h$$
$$\qquad (3.9)$$
$$u(t) = ak_e - ak_{12}/hk_{21}\exp(-k_{21}t)(1 - \exp(k_{21}h))$$
$$\text{if} \quad t \geqslant h$$

which gives the formula for explicit control as a function of h and k_{ij}. Its application to a concrete system associated with a drug developed by a pharmaceutical laboratory (Mrs Guerret - Labo-Sandoz) gives rise to the following numerical results. The drug is injected from time t = 0 and acts on the heart.

In (3.9) a is chosen equal to 18. Table 3.A gives the optimal control for five different values of h. We note that whatever the value of h, the control becomes constant from $t \geqslant t_0 \simeq 1.5$ h.

u(t)

t	h=0.19	h=0.15	h=0.10	h=0.05	h=0.03
0	94.736	120.000	180.000	350.000	600.000
0.01	97.804	123.885	185.827	371.655	619.424
0.03	103.492	131.090	196.636	393.271	55.4517
0.05	108.641	137.612	206.417	52.635	50.168
0.10	119.517	151.388	47.0816	41.328	39.3248
0.15	128.118	42.2825	37.006	32.684	31.179
0.19	39.009	33.401	30.7833	27.3454	27.345
0.20	37.206	31.910	29.4377	26.1910	25.060
0.50	12.0874	11.4036	10.6913	10.1079	9.9046
1.00	6.8990	6.8600	6.8194	6.7860	6.7740
1.50	6.6024	6.6003	6.5852	6.5851	6.5850
2.00	6.5850	6.5484	6.5844	6.5844	6.5844
2.50	6.5854	"	"	"	"
3.00	6.5854	"	"	"	"

Table 3.A

u(0) increases when h decreases. The following curves show the evolution of x_1 and u when h varies. Furthermore the exchange coefficients are as follows:

$k_{12} = 2.9545$, $K_{21} = 5.7214$, $k_e = 0.3658$

which gives a slow elimination of the drug.

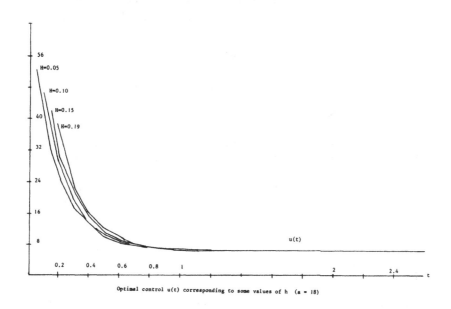

Optimal control u(t) corresponding to some values of h (a = 18)

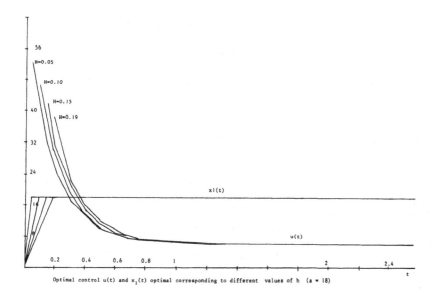

Optimal control u(t) and $x_1(t)$ optimal corresponding to different values of h (a = 18)

Another approach is possible by the use of convolution properties. We know that x_1 corresponding to u(t) is given by:

$$x_1(t) = x_1^*/\alpha * u(t) \tag{3.10}$$

where x_1^* are the observations and may be written as:

$$x_1(t) = A_1 exp(\lambda_1 t) + A_2 exp(\lambda_2 t) \tag{3.11}$$

with A_1, A_2, λ_1, λ_2 identified by means of an optimisation method. But $x_1^*(t)$ is a solution of a differential equation:

$$K(t) + \alpha K'(t) + \beta K''(t) = 0 \quad \text{with } K(t) = x_1^*/\alpha$$

where $\quad \alpha = k_e + k_{12} + k_{21} \quad$ and $\quad \beta = k_e k_{21}$

Consequently one can look for a solution of:

$$K(t) * u(t) = f(t) \tag{3.12}$$

where f(t) is fixed such that f(0) = 0. Differentiating (3.12) we obtain [62]:

$$[K(t) * u(t)]' = K'(t) * u(t) + K(0)u(t) = f'(t)$$

$$[K(t) * u(t)]' = K''(t) * u(t) + K'(0)u(t) + K(0)u'(t) = f''(t)$$

But $K + \alpha K' + \beta K'' = 0$ and therefore $(K + K' + K'') * u = 0$

that is to say:

$$\alpha K(0)u(t) + \beta K'(0)u(t) + \beta K(0)u'(t) = f(t) + f'(t) + f''(t) \tag{3.13}$$

u(t) is the solution of a linear differential equation (3.13) with constant coefficients and second term. An explicit solution can be obtained. An approximation of the previous situation may be realised by setting $f(t) \equiv a$ for all $t \geqslant 0$.

The optimal solution given by (3.13) is equal to:

$$u(t) = exp(-bt)u(0) + (a/\beta K(0)b)(1 - exp(-bt)) \tag{3.14}$$

with $b = (\alpha K(0) + \beta K'(0))/\beta K(0)$

In particular, if $u(0) = 0$, (3.14) implies:

$$u(t) = (a/\beta K(0)b)(1 - \exp(-bt))$$

Table 3.B corresponds to a concrete problem from pharmacokinetics (Mrs Guerret - Sandoz). The drug is a cardiac stimulant and two values of k_0 were tried. The values of other constants were:

$k_{12} = 1$, $K_{21} = 1.5$, $k_0 = 0.01$ or 0.1, $a = 10$ $\mu g/ml$. The given function is a concentration $C_1(t) = x_1/V_1$ where $V_1 = 4000$ ml. $u(t)$ must be replaced by $U(t) = u(t)/V_1$.

t	$u(t)$ for $k_0 = 0.01$	$u(t)$ for $k_0 = 0.1$
0	0	0
0.5	$70.888 \cdot 10^{-5}$	$55.945 \cdot 10^{-5}$
1.0	$90.870 \cdot 10^{-5}$	$64.311 \cdot 10^{-5}$
1.5	$96.503 \cdot 10^{-5}$	$65.562 \cdot 10^{-5}$
2.0	$98.090 \cdot 10^{-5}$	$65.749 \cdot 10^{-5}$
3.5	$98.699 \cdot 10^{-5}$	$65.782 \cdot 10^{-5}$
4.0	$98.709 \cdot 10^{-5}$	"
5.0	$98.713 \cdot 10^{-5}$	"
5.5	"	"
10.0	"	"
20.0	"	"

Table 3.B

3.3 The general solution [13]

The previous approach gives only one particular solution corresponding to a linear $x_1(t)$ satisfying $x_1(0) = 0$. Let us now examine what happens if we want $x_1(t) \equiv a$ for all $t \geq 0$. In this case our criterion:

$$\int_0^\infty (x_1(t) - a)^2 \, dt$$

reaches its absolute minimum value. Consider the differential system:

$$\dot{x}_1 = - (k_{12} + k_e)x_1 + k_{21}x_2 + u(t)$$

$$\dot{x}_2 = k_{12}x_1 - k_{21}x_2 \qquad\qquad (3.15)$$

$$x_1(0) = x_2(0) = 0$$

where $u(t) = a\delta_{(0)} + u_1(t)$ with $u_1(t)$ a classical function (not a distribution).

By Laplace transformation we observe that (3.15) is equivalent to the following:

$$\dot{x}_1 = - (k_{12} + k_e)x_1 + k_{21}x_2 + u_1(t)$$

$$\dot{x}_2 = k_{12}x_1 - k_{21}x_2 \qquad\qquad (3.16)$$

$$x_1(0) = a, \quad x_2(0) = 0$$

Indeed the Laplace transform gives the same transformed system for (3.15) and (3.16) because $\widehat{a\delta_{(0)}} = a.1 = a$.

But (3.16) allows (always using Laplace transform) an optimal solution realising $x_1(t) \equiv a$ for all $t \geq 0$. This solution is given by:

$$u_1(t) = a(k_e + k_{12} \exp(- k_{21}t)) \qquad\qquad (3.17)$$

Therefore the general solution of (3.15) is:

$$u(t) = a(\delta_{(0)} + k_e + k_{12} \exp(- k_{21}t)) \qquad (3.18)$$

where the first term corresponds to an instantaneous injection at time 0. Without this injection it is not possible to realise $x_1(t) \equiv a$ for all $t \geqslant 0$. In practice only the term ak_e is used; the term $ak_{12} \exp(- k_{21}t)$ is always neglected. Numerical results show that we cannot have $x_1(t) \equiv a$ in these circumstances. Furthermore if we try only $ak_e + ak_{12} \exp(-k_{21}t)$ we cannot have $x_1(t) \equiv a$ from the beginning.

3.4 Numerical method [36]

While looking for a function u(t) that minimises:

$$J(u) = \int_0^\infty (x_1(t) - a)^2 dt$$

with the constraints:

$$\dot{x}_1(t) = - (k_e + k_{12})x_1 + k_{21}x_2 + u(t)$$

$$\dot{x}_2 = k_{12}x_1 - k_{21}x_2$$

$$x_1(0) = x_2(0) = 0$$

where u(t) is a function, we showed that the solution was such that:

$$x_1(t) = K(t) \text{ } {}_{\star} \text{ } u(t) \qquad \text{with } K(t) = x_1^*/\alpha \qquad (3.19)$$

$x_1(t)$ is the measure associated with an injection α.

x_1^* and therefore K(t) is identified as:

$K(t) = A \exp(\lambda_1 t) + \exp(\lambda_2 t)$, $A, B, \lambda_1, \lambda_2$ known.

The initial minimisation problem may be made discrete by introducing a discretization of u(t) as follows:

$$u(t) = \sum_{p=1}^{n} c_p \theta_p(t) \qquad \text{(Ritz technique [49])} \qquad (3.20)$$

where the c_p parameters are unknown constants and $\theta_p(t)$ are known functions such as polynomials, exponentials, spline functions . . . . The initial problem becomes:

Find $c_1^*, \ldots, c_n^*$ such that:

$$J(c_1^*, \ldots, c_n^*) = \min_{c_1, \ldots, c_n} \int_0^\infty [K(t) * (\sum_{p=1}^n c_p \Theta_p(t)) - a]^2 dt \qquad (3.21)$$

In fact J is a second degree polynomial function in the unknowns c_p. The solution is obtained by writing:

$$\partial J/\partial c_p = 0 \quad , \quad p = 1, \ldots, n \qquad (3.22)$$

which is an algebraic linear system having n equations and n unknowns. More precisely:

$$\partial J/\partial c_p = 2 \int_0^\infty (K(t) * \Theta_p(t))(\sum_{i=1}^n c_i(K(t) * \Theta_i(t)) dt \qquad (3.23)$$

$$- 2a \int_0^\infty K(t) * \Theta_p(t) dt$$

The integrals are explicitly calculable if $\Theta_p(t)$ is an exponential, a polynomial or a spline function. It is easy to show that matrix A of system (3.23) is symmetric and therefore well established methods may be programmed.

Numerical experiments were performed with some particular spline functions [49]. Let N integer and h be such that $h = 1/N$ and choose $t_p = ph$. The Θ_p of formula (3.20) are as follows:

$$\Theta_p(t) = \begin{cases} 1 \text{ if } t = ph, \\ 0 \text{ if } t \notin [(p-1)h, (p+1)h], \\ -t/h + (p+1) \text{ if } t \in [ph,(p+1)h] \\ t/h - (p-1) \text{ if } t \in [(p-1)h, ph] \end{cases} \qquad (3.24)$$

We want to realise $x_1(t) \equiv 7$ for all $t \geqslant 0$ with:

$k_{12} = 0.16$, $K_{21} = 0.13$, $k_0 = 0.07$. The following tables (3.C to 3.F) give the values of c_p for n = 10, 15, 20 and the corresponding values of u(t) at various times t_i.

n=10 C_p

p	h = 0.5	h = 0.1	h = 0.05	h = 0.01
1	-8.501 E6	295512.0	155965.0	9302.62
2	2.961 E6	94026.4	49625.1	2648.17
3	-6.209 E6	80594.2	42535.2	- 6111.26
4	1.022 E7	-161188.0	- 85071.7	3530.99
5	-6.113 E6	-268647.0	-141786.0	1086.45
6	1.890 E6	53729.5	- 42535.3	- 4074.15
7	-5.349 E6	-537294.0	28357.3	-14123.8
8	-1.738 E7	-805594.1	-283572.0	-17111.5
9	1.566 E7	322377.0	170143.0	22136.4
10	2.559 E6	93228.6	49204.1	126523.0

Table 3.C

n=15 C_p in 10^6

p	h = 0.5	h = 0.1	h = 0.05	h = 0.01
1	4.04	9.03	4.77	-1.67
2	5.83	6.421	3.39	1.22
3	11.119	- 1.99	- 1.05	0.3334
4	11.119	4.39	2.31	1.49
5	4.081	- 8.13	- 4.29	-0.323
6	-17.70	- 9.17	0.455	-1.39
7	-77.77	0.863	- 3.44	1.72
8	-15.31	- 0.653	21.50	1.96
9	- 9.341	40.71	-10.70	0.983
10	-10.03	-20.20	7.14	1.98
11	- 9.34	13.50	-25.10	-1.39
12	50.20	- 0.926	- 0.488	7.41
13	- 6.02	0.726	4.053	6.49
14	-73.60	4.36	21.15	8.03
15	290.00	-35.41	-18.70	6.42

Table 3.D

n=20		c_p		
p	h = 0.5	h = 0.1	h = 0.05	h = 0.01
1	-8.40 E7	-3.36 E8	-1.77 E8	4.28 E6
2	-9.78 E7	-7.21 E7	-3.80	-2.89 E6
3	5.28 E7	4.66 E8	2,458 E8	7.75 E6
4	1.34 E8	-3.50 E8	-1.85 E8	-3.37 E6
5	-1.41 E8	4.27 E7	2.25 E7	-2.98 E6
6	-1.14 E8	2.42 E8	1.27 E8	8.71 E6
7	4.26 E7	-8.22 E7	-4.34 E7	1.32 E7
8	2.20 E7	-8.91 E7	-4.70 E7	-5.17 E6
9	-7.77 E7	1.12 E8	5.94 E7	8.06 E6
10	2.14 E8	-7.96 E8	-4.20 E8	-9.20 E6
11	3.16 E7	1.03 E8	5.44 E7	9.19 E6
12	-1.04 E8	-6.07 E8	-3.20 E8	1.39 E7
13	-2.19 E8	1.44 E8	7.61 E7	-2.28 E7
14	-3.73 E8	9.74 E7	5.14 E7	3.55 E7
15	1.33 E9	-4.11 E8	-2.17 E8	2.33 E8
16	7.74 E8	4.14 E8	2.18 E8	3.77 E7
17	6.56 E9	-1.49 E10	-7.85 E9	2.33 E8
18	-2.70 E8	-2.45 E9	-1.29 E9	1.89 E8
19	3.82 E8	-1.93 E9	-1.01 E9	-5.95 E7
20	1.04 E8	-7.23 E8	-3.81 E9	- 401905

Table 3.E

Numerical values of u(t) for h = 0.5

$t_i=ih$		$u(t_i)$	
i=0,..n	n = 10	n = 15	n = 20
0	8.5306 E7	3.111 E9	1.49667 E11
0.5	8.2745 E7	2.82269 E9	1.49563 E11
1	8.0186 E7	2.52869 E9	1.49459 E11
1.5	7.7627 E7	2.23469 E9	1.49251 E11
2	7.5068 E7	1.94069 E9	1.49251 E11
2.5	7.2509 E7	1.64669 E9	1.49147 E11
3	6.9950 E7	1.35269 E9	1.49043 E11
3.5	6.7391 E7	1.05869 E9	1.48939 E11
4	6.4832 E7	7.64686 E8	1.48835 E11
4.5	6.2273 E7	4.70686 E8	1.48731 E11
5	5.9714 E7	1.76686 E8	1.48627 E11
5.5		1.17314 E8	1.48523 E11
6		4.11314 E7	1.48419 E11
6.5		7.05314 E6	1.48315 E11
7		9.99314 E5	1.48211 E11
7.5		1.29331 E5	1.48107 E11
8			1.48003 E11
8.5			1.47899 E11
9			1.47795 E11
9.5			1.47691 E11
10			1.47587 E11

Table 3.F

 To complete this numerical study we may examine the influence of some coefficients k_{ij} on u(t) and $x_1(t)$. The following data is from the treatment of a two compartmental system (with initial injection) in Section 3.2.

$k_{12} = 2.9545$, $k_{21} = 5.7214$, $k_0 = 0.3658$, a = 18

If $x_1(0) = 18$ (corresponding to u(t) = $a\delta_{(0)}$) we obtain the following values of u(t) (without the term $a\delta_{(0)}$):

t	0	0.1	0.2	0.3	0.5	0.8
u(t)	59.7654	36.5953	23.5201	16.1415	9.62788	7.13134

t	1.0	1.2	1.5	1.8	2.0	2.5
u(t)	6.75875	6.63987	6.59437	6.58619	6.58497	6.58443

t	2.8	3.0	4.0	10	20	30
u(t)	6.58441	6.58444	6.58444	"	"	"

First case We study the $x_1(t)$ variation in the function of k_e^* occuring in u(t).   For that we set:

$$u^*(t) = ak_e^* + ak_{12}\exp(-k_{12}t)$$

and we calculate $\dot{x}_1^*(t)$ which is the solution of:

$$\dot{x}_1^*(t) = - (k_e + k_{12})x_1^*(t) + k_{21}x_2^*(t) + u^*(t)$$

$$\dot{x}_2^*(t) = k_{12}x_1^*(t) - k_{21}x_2^*(t) \qquad (3.25)$$

$$x_1^*(0) = a, \quad x_2^*(0) = 0$$

An explicit solution is calculated giving:

$$x_1^*(t) = a(k_e - k_e^*)((k_e + \lambda_2)/k_e\lambda_2 D + k_{12}/k_{21}k_e D)\exp(\lambda_1 t)$$

$$+ a/D(k_e - k_e^*)((k_e + \lambda_1)/k_e\lambda_1 + k_{12}/k_{21}k_0)\exp(\lambda_2 t)$$

$$+ ak_e^*/k_e$$

$$(3.26)$$

We can verify that if $k_e = k_e^*$ in u(t) then $x_1^*(t)$ given by (3.26) replaces $x_1(t)$.   Furthermore, if $k_e^* \to k_e$ we have:

$$x_1^* \to a \quad \text{and} \quad u^*(t) \to u(t)$$

The numerical results correspond to a real drug (Mrs Guerret - Sandoz) and take into account the data:

$k_{12} = 0.13, \quad k_{21} = 0.16, \quad k_e = 0.07, \quad V_1 = 4000 \text{ ml}$

V_1 is the volume of compartment 1.

k_e^* takes the following values 0.03, 0.05, 0.10, 0.20 and we want to have a = 7 μg/ml., where $C_1(t) = a = x_1/V_1$.

The following tables (Table 3.G and 3.H) and graphs give the numerical values of u*(t) and $C_1^*(t)$.

$$U^*(t) \text{ in } \mu g/ml$$

Time in hrs.	$k_e^*=0.03$	$k_e^*=0.05$	$k_e^*=0.07$	$k_e^*=0.10$	$k_e^*=0.20$
0	1.1200	1.2600	1.4000	1.61000	2.31000
0.1	1.1056	1.2456	1.3856	1.59556	2.29556
0.2	1.0913	1.2313	1.3713	1.58134	2.28134
0.5	1.0500	1.1900	1.3300	1.54004	2.24004
1.0	0.9854	1.12545	1.26545	1.47545	2.17545
1.5	0.92458	1.06583	1.2058	1.41583	2.11583
2.0	0.87079	1.01080	1.1508	1.36080	2.06080
3.0	0.77310	0.91310	1.0509	1.26309	1.96309
5.0	0.61880	0.75889	0.89889	1.10889	1.80889
10	0.39370	0.53370	0.67370	0.883726	1.58373
15	0.29255	0.43255	0.57255	0.782553	1.48255
20	0.2471	0.38709	0.521609	0.737094	1.43709
30	0.2175	0.35758	0.497429	0.707489	1.40749
40	0.2115	0.351512	0.491512	0.701512	1.40151
50	0.2105	0.35030	0.4903	0.700305	1.40031
60	0.2101	0.3501	0.4901	0.7001	1.4001
70	0.2100	0.3500	0.4900	0.7000	1.4000
80	0.2100	0.3500	0.4900	0.7000	1.4000
90	0.2100	0.3500	0.4900	0.7000	1.4000

Table 3.G

$$C_1^*(t) \text{ in } \mu g/ml$$

Time in hrs.	$k_e^*=0.03$	$k_e^*=0.05$	$k_e^*=0.07$	$k_e^*=0.10$	$k_e^*=0.20$
0	7	7	7	7	7
0.1	6.97166	6.98583	7	7.02126	7.09212
0.2	6.94265	6.97132	7	7.04302	7.18640
0.5	6.85209	6.92065	7	7.11093	7.48070
1.0	6.69177	6.64588	7	7.23117	8.00176
1.5	6.52341	6.76171	7	7.35744	8.54892
2.0	6.35059	6.67529	7	7.48706	9.11059
3.0	6.00244	6.50122	7	7.74817	10.2421
5.0	5.34491	6.17246	7	8.24132	12.3790
10	4.14260	5.57430	7	9.13855	16.2671
15	3.52534	5.26267	7	9.60599	18.2926
20	3.23293	5.11646	7	9.82531	19.2430
30	3.04408	5.04408	7	9.96694	19.98567
40	3.00820	5.00410	7	9.99385	19.9734
50	3.00152	5.00076	7	9.99886	19.9951
60	3.00010	5.00010	7	9.99980	19.9998
70	3	5	7	10	19.9999
80	3	5	7	10	20
90	3	5	7	10	20

Table 3.H

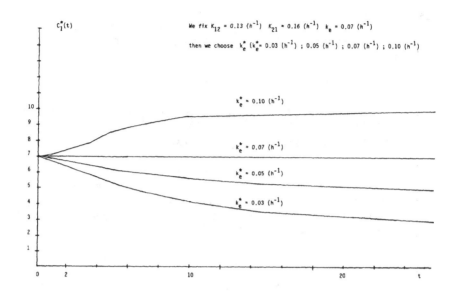

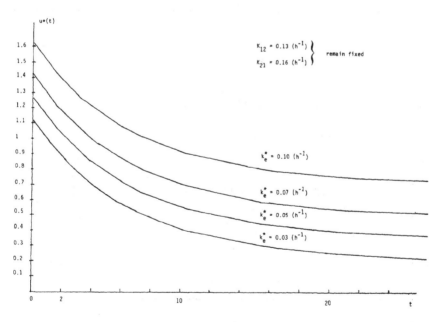

Second case. Here we study the variation of $x_1(t)$ in a function of k_{12}^* in $u(t)$. Proceeding, as before, to find an explicit relation, with $u^*(t) = ak_e + ak_{12}^*\exp(-k_{21}t)$, we obtain:

$$x_1^*(t) = a(k_{12}^* - k_{12})\exp(\lambda_1 t)/Dk_{21}$$

$$- a(k_{12}^* - k_{12})\exp(\lambda_2 t)/Dk_{21} + a \qquad (3.27)$$

where $D = (k_e + \lambda_2)/\lambda_2 - (k_e + \lambda_1)/\lambda_1$

The following results were obtained by fixing $k_{12} = 3.308$, $k_{21} = 1.4201$, $k_e = 1.175$ and k_{12}^* takes the values 3.308, 0.308, 0.0308, 9.924. a is equal to 15.

Table 3.I and the graphs which follow show the numerical values of $x_1^*(t)$ when k_{12}^* varies.

t	$k_{12}^*=0.03308$	$k_{12}^*=0.3308$	$k_{12}^*=3.308$	$k_{12}^*=9.924$
0	15	15	15	15
0.1	11.30400	11.6400	15	22.4668
0.5	7.61027	8.28206	15	29.5689
1.0	8.18576	8.80523	15	28.0148
1.5	9.09813	9.63467	15	26.9230
2.0	9.90973	10.3725	15	25.2834
2.5	10.6110	11.0100	15	23.8867
3.0	11.2157	11.5598	15	22.6450
5.0	12.9086	13.0987	15	19.2251
8.0	14.1407	14.2188	15	16.7359
10	14.5251	14.5683	15	15.9594
15	14.8922	14.9020	15	15.2179
20	14.9755	14.9777	15	15.0495
25	14.9944	14.9949	15	15.0112
30	14.9987	14.9989	15	15.0026
35	14.9997	14.9997	15	15.0006
40	14.9999	15	15	15.0001
50	15	15	15	15

Table 3.I

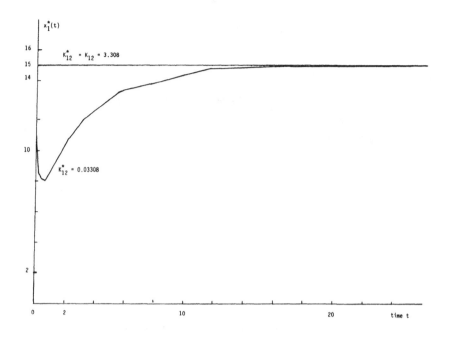

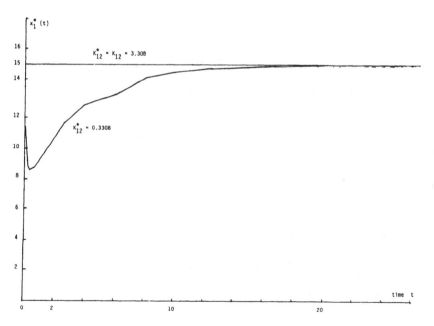

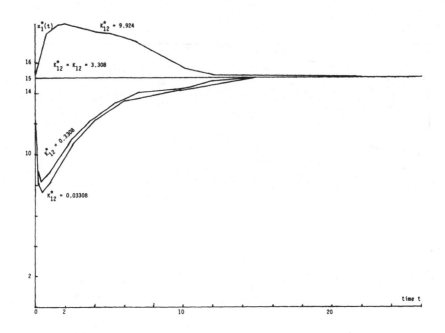

We observe that $x_1^*(t)$ goes from a = 15 to a minimum (or maximum) then increases (or decreases) to the constant value 15. Of course, for $k_{12}^* = k_{12} = 3.308$ we directly obtain $x_1^*(t) = x_1(t)$ a = 15.

Third case.   Here we study the evolution of $x_1^*(t)$, the solution of:

$$\dot{x}_1^*(t) = -(k_e^* + k_{12})x_1^*(t) + k_{21}x_2^*(t) + ak_e + ak_{12}\exp(-k_{21}t)$$

$$\dot{x}_2^*(t) = k_{12}x_1^*(t) - k_{21}x_2^*(t) \qquad\qquad (3.28)$$

$$x_1^*(0) = a, \quad x_2^*(0) = 0$$

when k_e^* varies (= 0.01, 0.03, 0.07, 0.5, 0.7).

An explicit resolution gives:

$$x_1^*(t) = k_1 \exp(\lambda_1 t) + k_2 \exp(\lambda_2 t) + a k_e/k_e^* \qquad (3.29)$$

with:

$$D = k_e^*(\lambda_1 - \lambda_2)/\lambda_1 \lambda_2$$

$$k_1 = - a(k_e^* + \lambda_2)(k_e^* - k_e)/\lambda_2 k_e^* D + a(k_e - k_e^*)k_{12}/k_{21}k_e^* D$$

$$k_2 = a k_{12}(k_e^* - k_e)/k_{21}k_e^* D + a(k_e^* + \lambda_1)(k_e^* - k_e)/\lambda_1 k_0^* D$$
$$\qquad (3.30)$$

The numerical applications used the following data:

$$k_{12} = 0.16, \quad K_{21} = 0.13, \quad k_0 = 0.07, \quad a = 7$$

and $\quad k_e^* = 0.01, 0.03, 0.07, 0.5, 0.7.$

The variation of $x_1^*(t)$ is described in Table 3.J and the graphs which follow.

t	$k_e^*=0.01$	$k_e^*=0.03$	$x_1^*(t)$ $k_e^*=0.07$	$k_e^*=0.5$	$k_e^*=0.7$
0	7	7	7	7	7
0.1	7.04164	7.02774	7	6.70871	6.57742
0.5	7.20150	7.13367	7	5.71703	5.20632
1.0	7.38755	7.25586	7	4.78921	4.03224
1.5	7.56025	7.36817	7	4.11309	3.25687
2.0	7.72139	7.47195	7	3.61560	2.73843
3.0	8.01506	7.65842	7	2.96516	2.14081
5.0	8.51964	7.97051	7	2.33542	1.65780
10	9.5383	8.57001	7	1.33542	1.24031
15	10.0797	9.06568	7	1.46432	1.02318
20	11.2742	9.51646	7	1.28370	0.893542
30	12.8993	10.3299	7	1.09947	0.769415
40	14.4527	11.0455	7	1.0270	0.724896
49	15.7933	11.495	7	1.0003	0.709893

Table 3.J

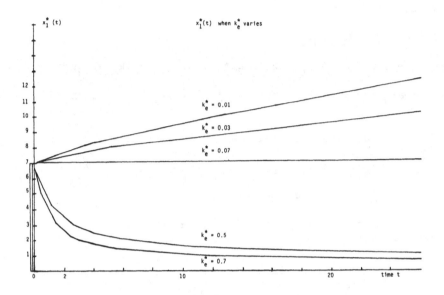

Fourth case This is an analog of (3.28) where the star (*)
is put on k_{12} instead of k_e.

$$\dot{x}_1^*(t) = - (k_e + k_{12}^*)x_1^*(t) + k_{21}x_2^*(t) + ak_e + ak_{12}\exp(-k_{21}t)$$

$$. \quad . \quad . \quad . \quad . \quad . \quad . \quad . \quad . \quad . \quad .$$

The reader may find x_1^* as a function of k_{ij}, k_{ij}^*.

3.5 Optimal control in non-linear cases [3]

3.5.1 A GENERAL TECHNIQUE

General control problems are defined using non-linear
differential systems. We want to minimise:

$$J(u) = \int_0^T f(x(t),u(t))dt \qquad (3.29)$$

where f is given, and knowing that $x(t)$ and $u(t)$ satisfy the equations:

$$\dot{x}(t) = g(x(t),u(t))$$

$$x(0) = x_0$$

(3.30)

where g and x_0 are given. Of course g may be non-linear.

To solve (3.29) and (3.30) discretization techniques associated with penalty methods can be used.

First:
$$\int_0^T f(x(t),u(t))dt$$

is approached by a formula approximating an integral (trapezoidal or rectangle rule, Gauss method, . .).

Taking, for example:

$$\int_0^T f(x(t),u(t))dt \simeq h \sum_{i=0}^{n-1} f(x_i,u_i)$$

(3.31)

where h is the discretization step ($nh = T$) and x_i, u_i are substituted for $x(t_i)$ and $u(t_i)$, $t_i = ih$.

(3.30) is discretized using Euler's method [24].

$$x_{i+1} = x_i + hg(x_i,u_i) = G(x_i,u_i)$$

$$x_1 \text{ known}$$

(3.32)

The initial problem (3.29) and (3.30) is transformed in a discretized optimization system:

$$\underset{u_0, \cdot \cdot \cdot ,u_{n-1}}{\text{Min}} \sum_{i=1}^{n-1} f(x_i,u_i) \text{ with constraints}$$

$$x_{i+1} = G(x_i,u_i) \quad , x_0 \text{ known } ; i = 0, 1, . . ,n-1$$

(3.33)

Introducing the penalty function:

$$J_\epsilon = \sum_{i=0}^{n-1} [f(x_i,u_i) + 1/\epsilon(x_{i+1} - G(x_i,u_i))^2] = \sum_{i=0}^{n-1} J_\epsilon^i$$

(3.34)

where $J_\epsilon^i = f(x_i,u_i) + 1/\epsilon(x_{i+1} - G(x_i,u_i))^2$

it can be proved that the minimum of J_ϵ as a function of x_1,
$\ldots,x_n,u_0,\ldots,u_{n-1}$ converges (when $\epsilon \to 0$) towards the
minimum of $\sum\limits_{i=1}^{n-1} f(x_i,u_i)$ according to $u_0,\ldots,u_{n-1}$ under the
constraints $x_{i+1} = G(x_i,u_i)$. For details one may consult
[18] or [49].
       Then the following algorithm is proposed:

The minimum of J_ϵ^i is looked for in terms of u_i. A control
u_i^ϵ, depending on ϵ, x_i, x_{i+1}, is obtained.
       Let us set:
$$u_i^\epsilon = r_\epsilon (x_i,x_{i+1}) \qquad (3.35)$$

From the structure of J_ϵ^i, the function r_ϵ depends only on ϵ
and not on i. To be more precise:

$$\underset{u_i}{\text{Min}} \quad J_\epsilon^i = J_\epsilon^i(x_i,x_{i+1},r_\epsilon(x_i,x_{i+1})) \qquad (3.36)$$

The numerical difficulty consists in determining r_ϵ for all
x_i, x_{i+1}. Using (3.32) we obtain:

$$u_i^\epsilon = r_\epsilon (x_i,G(x_i,u_i^\epsilon)) \qquad (3.37)$$

which allows the calculation of u_i, x_i, $i = 0,\ldots,n-1$
       Indeed we have:

$$u_0^\epsilon = r_\epsilon(x_0,G(x_0,u_0^\epsilon))$$

x_0 and G are known and r_ϵ calculated so that u_0^ϵ may be
calculated by solving an implicit equation.

       Then $u_1^\epsilon = r_\epsilon(x_1,G(x_1,u_1^\epsilon))$, $x_1 = G(x_0,u_0^\epsilon)$ implicitly
give u_1^ϵ and so on until u_n^ϵ, x_n. We now let ϵ go to zero.

       A first application follows. Consider the problem:

Find u(t) such that $J(u) = \int_0^T (x_1(t) -a)^2 dt$ be minimum
under constraints $\dot{x} = g(x,u)$, $x(0) = x_0$ where the control
u(t) interferes in a linear manner and with a single
component, u_1, not equal to zero, in compartment 1.

Therefore we may write:

$$\dot{x} = g(x,u) = g_1(x) + u$$

$$x(0) = x_0$$

(3.38)

This particular circumstance often arises in biological applications which can be modelled by non-linear compartmental systems. Furthermore we have:

$$f(x,u) = (x_1 - a)^2 \text{ ,where } x = (x_i), \quad i = 1, \ldots ,n.$$

It is easy to show that the previous technique converges, when $\epsilon \to 0$, towards the solution:

$$x_1(t) \equiv a \quad \text{for all } t \geqslant 0$$

$$u_1(t) = -g_{11}(x_1, \ldots ,x_n) = -g_{11}(a_1,x_2, \ldots ,x_n)$$

(3.39)

Indeed with (3.39) substituted in (3.38) we obtain a differential system with unknowns $x_2, \ldots , x_n$ and initial conditions $x_2(0) = x_{02}, \ldots ,x_n(0) = x_{0n}$ fixed.

In this case the control may be explicitly calculated provided that $x_1(0) = a$. If it were not the case, for instance, when $x_1(0) = 0$; $u(t)$ may be chosen as a distribution function.

$u(t) = a\delta_{(0)} + u_1(t)$ where $u_1(t)$ is equal to the solution given by (3.39).

<u>Remark</u> Of course, the method developed may be applied to differential systems of equations. To minimise the calculation one can use the Archimedean transformation to obtain functions of one variable. For example, identification of $r_\epsilon(x_i,x_{i+1})$ in (3.36) is possible using the transformation:

$$x_i = a\theta\cos\theta , \quad x_{i+1} = a\theta\sin\theta \quad \text{giving}$$

$$r_\epsilon(x_i,x_{i+1}) = r_\epsilon(a\theta\cos\theta,a\theta\sin\theta) = G_\epsilon(\theta)$$

3.5.2 A METHOD FOR NON-LINEAR COMPARTMENTAL SYSTEMS

Consider the following problem:

Minimise $\int_0^T (x_1(t) - a)^2 dt$ according to $u(t)$

where $u(t)$ and $x_1(t)$ satisfy the constraints
$$\dot{x} = g(x,u)$$
$$x(0) = x_0 \qquad\qquad (3.40)$$

with g and x_0 fixed and only the first component of $u(t)$ different from zero.

The technique is based on linearization of the differential system of (3.40). For this let us choose a discretization of $[0,T]$ that is to say $t_0 = 0$, t_1, . . . , $t_n = T$ in which a <u>linear</u> differential system may approach the non-linear one. This requires that the coefficients vary on each interval (t_i, t_{i+1}), $i = 0$, . . ,n-1. They are obtained using the Taylor expansion.

For example on $[t_0, t_1]$ we have the approximated linear system:
$$\dot{x} = A_0 x + b_0 u$$
$$\qquad\qquad (3.41)$$
$$x(0) = \alpha_0$$

A_0, b_0 calculated using Taylor's formula.

The minimization problem may be split as follows:

$$\underset{u(t)}{\text{Min}} \int_0^T (x_1(t) - a)^2 dt = \underset{u_0,\ldots,u_{n-1}}{\text{Min}} \sum_{i=0}^{n-1} \int_{t_i}^{t_{i+1}} (x_1(t) - a)^2 dt$$

where u_i is optimum on (t_i, t_{i+1}) and a sub-optimum is obtained using the algorithm:

$$\underset{u_0}{\text{Min}} \int_{t_0}^{t_1} (x_1(t) - a)^2 dt \qquad\qquad (3.42)$$

gives u_0 and the initial conditions on (t_1, t_2) solving:

$$\dot{x} = g(x,u), \quad x(0) = \alpha_0$$

$$\text{Min}_{u_1} \int_{t_1}^{t_2} (x_1(t) - a)^2 dt$$

gives u_1 and the initial conditions on (t_2, t_3) by resolving $\dot{x} = g(x,u)$, $x(0) = \alpha_1$ and so on until:

$$\text{Min}_{u_{n-1}} \int_{t_{n-1}}^{t_n} (x_1(t) - a)^2 dt$$

The minimization:

$$\int_{t_i}^{t_{i+1}} (x_1(t) - a)^2 dt$$

according to u_i is solved explicitly because the system:

$$\dot{x} = A_i x + b_i u$$

$$x(0) = \alpha_i$$

is of the same type as those tested previously. The Laplace transformation gives an explicit solution which is a combination of a constant term and of $(n - 1)$ exponential terms using the parameters of k_{ij} of the matrix A_i. For more details see chapter 4.

It is clear that applying this technique gives only a local optimum and not the global optimum. In practice, this sub-optimum is often sufficient and gives a better solution than the empirical one found by experiment.

Furthermore when the discretization step goes to zero the solution $x_1(t)$ on each interval (t_1, t_{i+1}) tends towards a and the optimal solution $u^*(t)$ goes to the absolute minimum of:

$$\int_0^T (x_1(t) - a)^2 dt.$$

Remark The reader may easily generalise these results to more general situations where the state system:

$$\dot{x} = P(x,u)$$

is also non-linear according to the control $u(t)$ and where

the criterion is defined by a non-linear function such as
the following:

$$\underset{u(t)}{\text{Min}} \quad \int_0^T G(x,u)dt$$

Only one of these situations need be considered.
Decomposing $[0,T]$ in sub-intervals (t_i, t_{i+1}) where G is
linearised with respect to u gives the sequence of
linearization problems:

$$\underset{u(t)}{\text{Min}} \quad \int_{t_i}^{t_{i+1}} [G(x_i,u_i) + (u - u_i)G_u'(x_i,u_i)]dt$$

This technique will give a sub-optimum, as does the
previous one. Then the differential equation $\dot{x} = P(x,u)$ is
also linearised with respect to x and u using a first order
Taylor expansion giving a linear differential system:

$$\dot{x} = P(x_i,u_i) + (u - u_i)P_u'(x_i,u_i) + (x - x_i)P_x'(x_i,u_i)$$

As for the linear case, the use of the Laplace
transformation allows an exact optimal solution u(t) on each
interval (t_i, t_{i+1}) to be found. Nevertheless this method
fails if the differential system is stiff [76].

3.5.3 ANOTHER PRACTICAL APPROACH

In general, when studying the previous problem (3.40) we had
at our disposal an $x_1^*(t)$ corresponding to each of various
doses of $\alpha_1, \ldots, \alpha_m$.

An approximation of $x_1^*(t)$ may be tried in the shape of:

$$x_1^*(t) = \sum_{i=1}^{n} a_i \exp(\lambda_i t) \qquad (3.43)$$

where the a_i and λ_i have to be identified as functions of α.
An optimization technique will be needed. Therefore the
functions $a_i(\alpha)$ and $\lambda_i(\alpha)$ will be obtained. Then a first
optimal problem which consists of finding the doses α_i and
times of injection t_i such that $x_1(t)$ be as near as possible

to a constant a may be solved. Mathematically the problem
is as follows:

Find doses α_i and times t_i such that:

$$\int_0^T (x_1(t) - a)^2 dt \qquad \text{be minimum (T fixed)} \qquad (3.44)$$

It is necessary to define $x_1(t)$ when t varies because at
each dose α_i given at time t_i there is a corresponding
function $x_1^i(t)$. Suppose that we are working with
concentrations x_1, x_1^i and with constant volume V_1. We may
admit the <u>additivity</u> of concentrations during the time. At
(t_i, t_{i+1}) the following concentration is measured:

$$x_1^i(t) = \sum_{j=0}^{i} x_1^j(t) = \sum_{j=0}^{i} [\sum_{k=1}^{n} a_k(\alpha_j) \exp(\lambda_k(\alpha_j)(t-t_j))] \qquad (3.45)$$

The optimal control problem becomes:

$$\underset{\substack{\alpha_i, t_i \\ i=0,\ldots,m-1}}{\text{Min}} \sum_{i=0}^{m-1} \int_{t_i}^{t_{i+1}} [\sum_{j=0}^{i} (\sum_{k=1}^{n} a_k(\alpha_j) \exp(\lambda_k(\alpha_j)(t-t_j))) - a]^2 dt \qquad (3.46)$$

which is a classical minimization problem. The numerical
solution may be found using methods in this chapter.
Generalization is possible considering that a general
control function u(t) may be approached by:

$$u(t) = \sum_{i=0}^{m-1} \alpha_i \delta_{(t_i)} \qquad (3.47)$$

where $\delta_{(t_i)}$ is the Dirac delta-function at time t_i [59]. If

we prefer classical functions the delta-function can be
approximated by step functions and the formula (3.47)
becomes:

$$u(t) = \sum_{i=0}^{m-1} \alpha_i \theta_i(t) \qquad (3.47a)$$

where $\Theta_i(t)$ is the characteristic function of the interval $(ih,(i+1)h)$. (3.47a) is a general approximation for a function $u(t) \in L^2(0,T)$. Therefore our simplified approach can solve optimal control problems where a general control function $u(t)$ is required. The optimal problem associated with (3.47a) is exactly the same as (3.47).

A slightly different technique is possible if we can make use of the following data: $x_1(t)$ measured for different input functions $u(t)$. The next integral formula may be tried:

$$x_1(t) = K * g(u) \tag{3.48}$$

where K is known and corresponds to a linear compartmental system. The identification of K is possible using data (u, x_1) and writing $x_1 = K * u$. Then if g is chosen as a polynomial function:

$$g(u) = \sum_{p=0}^{r} a_p u^p$$

the coefficients a_p are identified by minimising J with:

$$J = \sum_{j=1}^{m} [x_1^{(j)}(t) - K * \sum_{p=0}^{r} a_p u_j^p]^2 \tag{3.49}$$

The solution is given by solving the linear algebraic system:

$$\partial J/\partial a_p = 0, \quad p = 0, \ldots, r \tag{3.50}$$

The optimal control problem:

$$\underset{u}{\text{Min}} \int_0^T (x_1(t) - a)^2 dt = \underset{u}{\text{Min}} \int_0^T (K * g(u) - a)^2 dt$$

becomes a classical optimization problem.

Remark Previously the following relation was shown:

$$x_1(t) = x_1^*/\alpha * u = K * U \tag{3.51}$$

where K corresponds to a Dirac delta-function at time zero. When the system is non-linear, the relation (3.51) is modified as follows:

$$x_1(t) = K * g(u) \tag{3.52}$$

with $g(u)$ non-linear and to be identified in a particular form (polynomial, . . .).

3.5.3 A VARIANT OF DYNAMIC PROGRAMMING TECHNIQUE [5], [11]

The dynamic programming technique was first developed by R. Bellman [5]. Applications to biomedical problems are also proposed by this author in [4]. Our technique is based on the discretization proposed earlier. (See section 3.5.2). We saw that the discretized problem was:

$$\text{Find } u_0, \cdots u_{n-1} \text{ minimising } \sum_{i=0}^{n-1} f(x_i, u_i)$$

under constraints $x_{i+1} = G(x_i, u_i)$, $i = 0, \cdots, n-1$.
x_0 known. $\tag{3.53}$

Let us write the fundamental relation of dynamic programming

$$\underset{u_0, \cdots, u_{n-1}}{\text{Min}} \sum_{i=1}^{n-1} f(x_i, u_i) = \underset{u_0, \cdots, u_{n-2}}{\text{Min}} [\sum_{i=1}^{n-2} f(x_i, u_i)$$

$$+ \underset{u_{n-1}}{\text{Min}} f(x_{n-1}, u_{n-1})] \tag{3.54}$$

This equality is valid because only $f(x_{n-1}, u_{n-1})$ depends on u_{n-1}.

Setting:

$$\underset{u_{n-1}}{\text{Min}} f(x_{n-1}, u_{n-1}) = G_{n-1}(x_{n-1}) \tag{3.55}$$

where G_{n-1} can be calculated numerically using an optimization technique. Notice that G_{n-1} must be calculable for all x_{n-1} and therefore a particular structure of G_n has to be identified. For example, one may choose a polynomial or an exponential function. The optimal u_{n-1} of (3.55) will be denoted by $u_{n-1}^* = \varphi_{n-1}(x_{n-1})$ where φ_{n-1} may be identified by a numerical optimization technique.

The next step is to solve:

$$\underset{u_{n-2}}{\text{Min}} \quad [f(x_{n-2}, u_{n-2}) + G_{n-1}(x_{n-1})] \qquad (3.56)$$

But from (3.53) $x_{n-1} = G(x_{n-2}, u_{n-2})$ and therefore:

$$\underset{u_{n-2}}{\text{Min}} \quad [f(x_{n-2}, u_{n-2}) + G_{n-1}(G(x_{n-2}, u_{n-2}))]$$

has to be solved. The functional to be minimised depends only on x_{n-2} and u_{n-2}. As before, we take the two functions:

$$G_{n-2}(x_{n-2}) = \underset{u_{n-2}}{\text{Min}} \quad [f(x_{n-2}, u_{n-2}) + G_{n-1}(G(x_{n-2}, u_{n-2}))]$$

$$\qquad (3.57)$$

$$u_{n-2}^* = \varphi_{n-2}(x_{n-2})$$

Then the problem arises:

$$\underset{u_{n-3}}{\text{Min}} \quad [f(x_{n-3}, u_{n-3}) + G_{n-2}(x_{n-2})]$$

In the same way the functions:

(G_{n-1}, φ_{n-1}), (G_{n-2}, φ_{n-2}), . . . , (G_1, φ_1), (G_0, φ_0) are calculated. But $u_0^* = \varphi_0(x_0)$ with x_0 known allows the determination of u_0^*. Then $x_1^* = G(x_0, u_0^*)$ gives $u_1^* = \varphi_1(x_1^*)$, and so on. Therefore the u_0^*, x_1^*, u_1^*, x_2^*, u_2^*, x_3^*, . . . , x_{n-1}^*, u_{n-1}^*, x_n^*, are determined as required.

In contrast to the preceding method this technique converges towards the global minimum. For this it is necessary to calculate the values u_{n-1}^*, u_{n-2}^* and so on, defining the global minimum.

For example, if we apply this method to a particular differential system: $\dot{x} = g(x, u) = g_1(x) + u$ we obtain:

$$f(x_i, u_i) = (x_{1i} - a)^2 = [x_{1i+1} - h(g_{11}(x_i) + u_i) - a]^2$$

from $x_{1i+1} = x_{1i} + h(g_{11}(x_i) + u_i)$

The minimum of $f(x_i, u_i)$ as a function of u_i is given by:

$$\partial f(x_i, u_i)/\partial u_i = 0 \iff u_i = x_{1i+1}/h - g_{11}(x_i) - a/h$$

$$i = 0, 1, \ldots, n-1 \tag{3.58}$$

We deduce:

$$\varphi_i(x_{1i}) = x_{1i+1}/h - g_{11}(x_i) - a/h, \quad i = 1, \ldots, n-1$$

From:

$$x_{1i+1} = x_{1i} + h(g_{11}(x_i) + u_i)$$

and (3.58) we further deduce:

$$x_{1i} = a \text{ for } i = 1, \ldots, n-1$$

Finally: $\quad u_i = -g_{11}(x_i)$, that is, $\quad \varphi_i(x_{1i}) = -g_{11}(x_i)$

Then u_0, x_1, u_1, x_2, $\cdot$ $\cdot$, u_{n-1}, x_n are successively determined as follows:

$$u_0^* = - g_{11}(x_0)$$

$$x_{11} = x_{10} + h(g_{11}(x_0) + u_0^*)$$

$$u_1^* = - g_{11}(x_1) \tag{3.59}$$

$$x_{12} = x_{11} + h(g_{11}(x_1) + u_1^*)$$

$$\cdot$$
$$\cdot$$
$$\cdot$$

This method can be applied to the non-linear model shown in Figure 10.

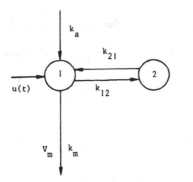

This model, studied in [72], is represented by the differential system:

$$\dot{x}_1 = k_a x_0 \exp(-k_a t) - (k_{12} + V_m/(K_m + x_1(t)))x_1(t) + k_{21}x_2(t) + u(t)$$

$$\dot{x}_2 = k_{12}x_1(t) - k_{21}x_2(t) \tag{3.60}$$

$$x_1(0) = x_{10}, \quad x_2(0) = x_{20}$$

The associated control problem is to maintain, as far as possible, $x_1(t) \equiv a$. Mathematically, this is equivalent to:

$$\underset{u(t)}{\text{Minimize}} \quad \int_0^T (x_1(t) - a)^2 dt \tag{3.61}$$

where x_1, x_2, u are related by equations (3.60)

Our method is applied with $a = 7$ and the following numerical values:

$$k_{12} = 3, \quad k_{21} = 2, \quad k_a = 2, \quad x_0 = 2$$

$$V_m = 0.418, \quad K_m = 0.182, \quad x_2(0) = 0$$

Two cases are considered for $x_1(0)$, namely $x_1(0) = 0$ and $x_1(0) = 7$. Two values of h (0.1 and 0.3) are used. Numerical results for u_i^*, x_{1i}, x_{2i} are shown in the following tables (Tables 3.K and 3.L) and graphs.

$$h = 0.1 \quad x_2(0) = 0 \quad n = 500$$

$$x_1(0) = a = 7 \qquad\qquad x_1(0) = 0$$

t_i	u_i	x_{1i}	x_{2i}	u_i	x_{1i}	x_{2i}
0	17.4074	7	0	– 4	0	0
0.1	13.9325	7	2.1	18.1324	7	0
0.2	11.1662	7	3.78	14.526	7	2.1
0.3	8.9642	7	5.124	11.652	7	3.78
0.5	5.8172	7	7.05936	7.53748	7	6.124
0.8	3.1230	7	8.73839	4.0038	7	8.29799
1.0	2.1209	7	9.37257	2.6846	7	9.09070
1.2	1.4876	7	9.778844	1.8484	7	9.59805
1.5	0.94713	7	10.13056	1.1318	7	10.038205
2.0	0.57625	7	10.3789	0.63678	7	10.34867
2.5	0.45979	7	10.4600	0.47963	7	10.4504
3.0	0.42349	7	10.4870	0.42998	7	10.4837
5.0	0.407525	7	10.4998	0.40760	7	10.4999
8.0	0.407407	7	10.4999	0.407407	7	10.4999
10	0.407407	7	10.5	0.407407	7	10.5
20	0.407407	7	10.5	0.407407	7	10.5
30	0.407407	7	10.5	0.407407	7	10.5
40	0.407407	7	10.5	0.407407	7	10.5
50	0.407407	7	10.5	0.407407	7	10.5

Table 3.K

$h = 0.3 \quad x_1(0) = a = 7 \quad x_2(0) = 0 \quad n = 125$

t_i	u_i	x_{1i}	x_{2i}
0	10.4074074	7	0
0.3	6.61216	7	6.3
0.9	1.0902	7	9.828
1.5	0.423299	7	10.3925
2.1	0.384128	7	10.4829
2.7	0.394846	7	10.497
3.0	0.39969	7	10.497
3.6	0.40477	7	10.4988
4.2	0.40656	7	10.4999
5.1	0.40726	7	10.4999
6.0	0.40738	7	10.4999
8.1	0.407406	7	10.5
10.2	0.4074074	7	10.5
12.0	0.4074074	7	10.5
15.0	0.4074074	7	10.5
20.1	0.4074074	7	10.5
30.0	0.4074074	7	10.5
30.9	0.4074074	7	10.5
33.0	0.4074074	7	10.5
33.1	0.4074074	7	10.5
37.5	0.4074074	7	10.5

Table 3.L

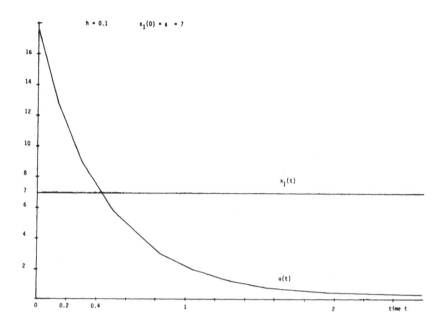

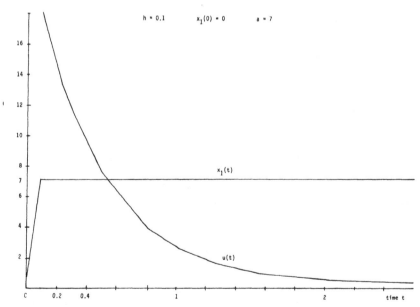

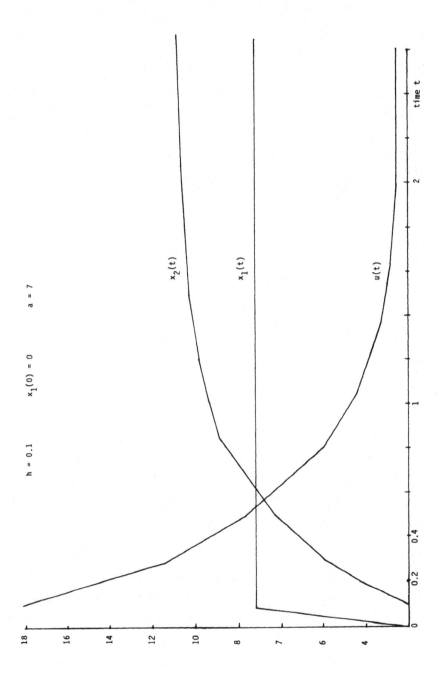

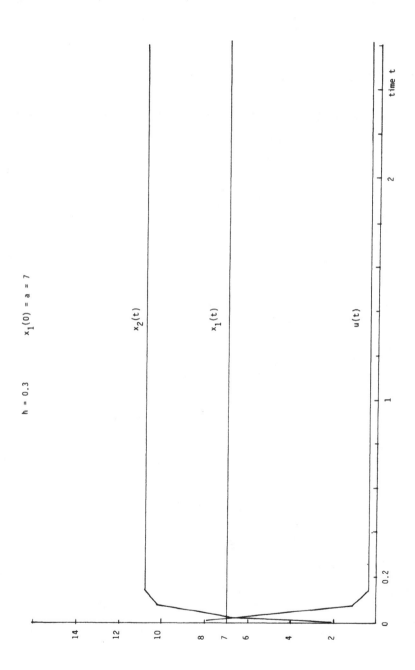

3.5.4 A simple idea applied to optimal control problems

Once again, consider the general control problem:

Minimise $\qquad\qquad\qquad \displaystyle\int_0^T f(x(t),u(t))dt$

where it is known that x and u satisfy the differential system:

$$\dot{x}(t) = g(x(t),u(t))$$

$$\hspace{6cm}(3.63)$$

$$x(0) = x_0$$

An approximation of the integral in (3.62) leads to:

$$\int_0^T f(x(t),u(t))dt \simeq h \sum_{i=1}^{n} g(x_i,u_i,t_i) \qquad (3.64)$$

by use of a quadrature formula. In (3.64), h is the discretization step and we set $x_i = x(t_i)$, $u_i = u(t_i)$.

Then:

$$\underset{u(t)}{\text{Min}} \int_0^T f(x(t),u(t))dt \simeq \underset{u_1,\ldots,u_n}{\text{Min}} \; h \sum f(x_i,u_i,t_i)$$

$$= h \sum_{i=1}^{n} \underset{u_i}{\text{Min}} \; f(x_i,u_i,t_i)$$

It is sufficient to determine the function $F(x_i,t_i)$ such that:

$$\underset{u_i}{\text{Min}} \; f(x_i,u_i,t_i) = f(x_i,u_i^*,t_i)$$

$$\hspace{6cm}(3.65)$$

$$\text{with} \quad u_i^* = F(x_i,t_i)$$

Note that the function F does not depend on i. The Alienor transformation may be used to identify F. It reduces the problem to the determination of a function of a single variable θ by setting $x_i = a\theta\cos\theta$, $t_i = a\theta\sin\theta$.

The generalization to a vector x(t) presents no

difficulty. In that case the tree structure of the Alienor transformation has to be employed.

Putting the relation:

$$u(t) = F(x,t) \qquad (3.66)$$

into the differential system (3.63) gives the optimal state variable (or vector) $x^*(t)$. Optimal control is obtained from the relation:

$$u^*(t) = F(x^*(t),t) \qquad (3.67)$$

This method avoids many difficulties associated with classical methods such as the dynamic programming one, since it requires fewer calculations and less computer memory.

CHAPTER 4

RELATIONS BETWEEN DOSE AND EFFECT

4.1 General considerations

The pharmacologist using a drug is interested initially in
its action and then in the optimization of this action. It
is sometimes easy to model the action, especially when there
is a direct and linear relation to the concentration in the
blood compartment [72]. In many cases the action of the
drug cannot be related to the concentration in the blood by
a simple model. This happens when sophisticated biochemical
transformations arise [35]. In this chapter we shall
describe some of the main difficulties.

First, let us begin with the simplest models.

a) Compartmental models

We suppose that the drug conforms to a linear compartmental
model where one of the compartments is considered as the
"deep compartment" or the reaction compartment. In this
case it is not difficult to find the relation between the
blood concentration and the action. Let us consider the
compartmental model in Figure 11.

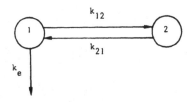

FIG. 11

The equations have already been shown:

$$\dot{x}_1 = - (k_{12} + k_e)x_1 + k_{21}x_2$$

$$\dot{x}_2 = k_{12}x_1 - k_{21}x_2 \qquad (4.1)$$

$$x_1(0) = \alpha, \quad x_2(0) = 0, \quad \alpha \text{ known}$$

96

Finding a relation between the blood concentration $x_1(t)$ and the drug action $x_2(t)$ is not difficult. The use of the Laplace transformation gives:

$$s\hat{x}_1 - \alpha = -(k_{12} + k_e)\hat{x}_1 + k_{21}\hat{x}_2$$
$$s\hat{x}_2 = k_{12}\hat{x}_1 - k_{21}\hat{x}_2 \tag{4.2}$$

using the relation:

$$\hat{x}_2 = k_{12}\hat{x}_1/(s + k_{21}) \tag{4.3}$$

Returning to the original we have:

$$x_2(t) = k_{12}x_1 * (\exp(-k_2 t))Y(t) \tag{4.4}$$

where $Y(t)$ is the Heaviside function equal to zero for $t < 0$ and equal to 1 for $t \geq 0$.

In the general case of an n-compartment model the relation between the blood concentration x_1 and the action $x_i(t)$ is given by a convolution:

$$x_i(t) = K(t) * x_1(t) \tag{4.5}$$

where $K(t)$ may be explicitly calculated. In fact $K(t)$ is a linear combination of exponential functions.

Remark The effect of a drug depends on its chemical nature. In fact any effect may be taken into consideration provided it is quantified. For example, an effect may be described by records of arterial pressure, heart rate, concentration of the drug in one organ, electrical activity, . . . [28].

When the method adopted is not compartmental analysis, the following approach is used:

b) Linear relationships

The classical literature [72] generally agrees that from practical considerations the effect is a linear function of the logarithm of the concentration. In other words, we have

$$E(t) = a_0 + a_1 \log(C(t)) \tag{4.6}$$

where $E(t)$ is the effect during the time, and $C(t)$ the concentration of the drug in the blood. The coefficients a_0

and a_1 are unknown and have to be found by solving the optimization problem:

$$\underset{a_0, a_1}{\text{Min}} \sum_{j=1}^{m} [E(t_j) - a_0 - a_1 \log(C(t_j))]^2 = \underset{a_0, a_1}{\text{Min}} \quad J \qquad (4.7)$$

where the t_j are the times of measurement of C and E. The problem (4.7) may be reduced to a linear algebraic system by writing:

$$\partial J/\partial a_0 = \partial J/\partial a_1 = 0$$

This technique has been used in the previous chapters. Relating dose and effect is therefore easy when this approach is possible. If the relation is non-linear there may be difficulties which we shall now discuss.

4.2 The non-linear approach

We are looking for a general relation, which can be simulated, between the effect $x_2(t)$ and the concentration $x_1(t)$.

The formula:

$$x_2 = F(x_1(t), D) \qquad (4.8)$$

is a mathematical relation which links x_1, x_2 and the oral dose D given at time t = 0. A very general problem arises where an oral dose is given. Problems with injections (Dirac delta-function) at time t = 0 are easier to solve. As usual it is best to try the simplest formula first (4.8).

For instance we may begin with:

$$x_2(t) = b_0 + b_1 \log(1 + x_1(t)) \qquad (4.9)$$

where the constant 1 is included to ensure that the logarithmic function is positive. As in Section 4.1 a linear regression may be used to identify the two constants b_0, b_1. If the approximation is poor a more sophisticated formula (4.8) must be tried. As suggested by the Weber-Fechner law, which states that sensation is approximately proportional to the logarithm of the stimulus, one may try:

$$x_2(t) = b_0 + b_1 \log(1 + x_1(t)) + b_2 \log^2(1 + x_1(t)) \qquad (4.10)$$

where a non-linearity is introduced due to the term $\log^2$. The coefficients b_0, b_1, b_2 have to be identified from experimental data using a regression technique as before.

To test this theory we show a table which corresponds to a beta-blocker studied by Dr. J.F. Prost in a French laboratory (Labo. Servier). Four doses were tested on ten subjects during exercise. Measurements were made at different times t_i. $x_2(t)$ is the measured value of effect, x_2^c the calculated value of x_2 and $x_1(t_i)$ the measured concentration.

The experimental data shown in Table 4.A are the mean values obtained from the ten subjects. The numerical results were obtained by A. Guillez (Medimat) using linear regression and the formula:

$$x_2 = b_0 + b_1 \log(n + x_1(t)) + b_2 \log^2(n + x_2(t)) \quad (4.11)$$

The best approximation is found when $n = 0.1$ but there is little improvement over $n = 1$. In (4.11) b_0 was calculated, but this is not necessary since one may set $b_0 = x_2(0)$ and then only b_1 and b_2 need be calculated. The previous table shows that b_1, b_2 depend on the dose D. An explicit relation may be proposed by using an optimization technique. The following formulae are obtained:

$$b_1 = 01.454D/(D + 0.651)$$

$$b_2 = 0.0278 - 0.1638/\sqrt{D}$$

Dose	1 mg			2 mg			5 mg			10 mg		
Hrs.	$x_1(t)$	$x_2(t)$	$x_2^c(t)$	$x_1(t)$	$x_2(t)$	$x_2^c(t)$	$x_1(t)$	$x_2(t)$	$x_2^c(t)$	$x_1(t)$	$x_2(t)$	$x_2^c(t)$
0	0.0	28.5	27.0	0.0	26.1	25.7	0.0	24.5	24.5	0.0	25.4	25.4
0.5	7.7	25.5	23.0	28.9	21.3	18.7	58.5	19.0	16.2	15.0	17.0	14.9
1.0	15.6	22.6	21.9	32.3	18.7	18.5	88.9	15.9	15.6	190.4	14.2	14.5
2.0	20.0	20.0	21.5	32.5	17.0	18.4	111.4	14.5	15.2	197.9	13.8	14.5
4.0	11.1	21.2	22.4	22.1	17.3	19.1	71.6	14.7	15.9	116.0	14.4	15.3
6.0	5.7	23.9	23.4	11.7	20.5	20.2	38.2	26.1	16.9	66.5	16.4	16.2
9.0	1.4	24.6	25.1	5.2	21.3	21.4	20.7	17.7	17.8	33.5	16.3	17.3
12.0	0.3	25.5	26.3	2.6	22.6	22.4	10.9	18.1	18.7	20.5	17.9	18.0
15.0	0.15	25.9	26.6	0.8	24.8	23.7	5.2	20.3	19.7	11.7	19.4	18.8
24.0	0.0	26.4	27.0	0.3	23.2	24.6	2.0	20.8	21.0	5.2	20.1	20.0
b_0	25.5			23.62			21.89			22.38		
b_1	-0.954			-1.143			-1.218			-1.38		
b_2	-0.131			-0.0979			-0.0418			-0.02265		

Table 4.A

Remark Another possible approach is to use a convolution
relation as below:

$$x_2(t) = x_2(0) + \int_0^t K(t - \tau)x_1(\tau)d\tau \qquad (4.12)$$

Equation (4.12) is a direct generalization of a linear
compartmental model, but it allows non-linear models to be
considered easily, if necessary. To solve (4.12) a well
known numerical method is necessary. Using an approximation
for $x_1(t)$:

$$x_1(t) = \sum_{i=1}^3 a_i \exp(\lambda_i t) \qquad (4.13)$$

where a_i and λ_i are determined by preceding techniques.
$K(t)$ is calculated as a polynomial function:

$$K(t) = b_0 + b_1 t + b_2 t^2 + b_3 t^3 \qquad (4.14)$$

Substituting (4.13) and (4.14) in (4.12) we obtain:

$$x_2(t) = \sum_{j=0}^3 b_j \theta_j(t) \qquad (4.15)$$

where $\theta_j(t)$ is a function calculated from the a_i and λ_i of
(4.13) The b_j are calculated by minimising the functional:

$$J = \sum_{k=1}^m [x_2(t_k) - \sum_{j=0}^3 b_j \theta_j(t_k)]^2$$

with t_k, $k = 1, \ldots, m$ times of measurement.

Numerical simulation does not give a good approximation. It
confirms that, in some cases, linear models are not
consistent with experimental results. So a non-linear
integral equation may be tried:

$$x_2(t) = x_2(0) + \int_0^t K(t - \tau)[x_1(\tau) + \alpha x_1^2(\tau)]d\tau \qquad (4.16)$$

Numerical experiments (see Table 4.B, below) show that

it is possible to make a good approximation to the concentration of drug in the blood (x_1) and the effect x_2, but the function K depends on the dose D. The table is associated with D = 2 mg. In the table, t = time in hours, x_2 is experimental, and x_2^c is calculated.

t	0	0.5	1	2	4	6	9	12	15	24
x_2 exp	26.1	21.3	18.8	17.1	17.4	20.5	21.3	22.7	24.8	23.3
x_2^c calc $\alpha=0.2$	26.1	20.3	19.4	17.9	17.5	19.2	21.8	23.5	24.2	23.3

Table 4.B

Although this non-linear model is well adapted to the experimental data, difficulties remain in identifying K as a function of D. Further improvements are possible using another simplified technique.

4.3 Simple functional model

We come back to the general relation:

$$x_2(t) = F(x_1(t), D) \tag{4.17}$$

The shape of the curve $x_2(t)$, $t \geqslant 0$, suggests, for F

(i) a straight line on $0 \leqslant t \leqslant 1$
(ii) a constant added to an exponential on $t \geqslant 1$

Using an optimization technique to identify these functions leads to:

$x_2(t) = -8.45t + b_0$ if $0 \leqslant t \leqslant 1$
where b_0 is equal to $x_2(0)$ given

$x_2(t) = b_0 + a(D)\exp(\lambda(D)t)$ if $t \geqslant 1$ $(\lambda(D) < 0)$
where b_0 is the same as before,
       since $x_2(t) \to b_0$ when $t \to +\infty$

$$\tag{4.18}$$

It can be seen that a and λ depend on D, but the experimental data, on the interval $0 \leqslant t \leqslant 1$, involves constant coefficients (independent of D). Using identification methods we obtain:

$$a(D) = (-13.6D - 14.38)/(D + 2.53)$$

$$\lambda(D) = (-0.038D + 0.43)/(D - 1.38)$$

(4.19)

The main inconvenience of this approach is that we obtain a function F which is not defined by a unique expression on all $t \geqslant 1$. Note that these results confirm that models associated with a biological phenomenon are not unique. We choose the model which is the simplest for mathematical studies. The simplicity is often subjective and depends on the experience of the biomathematician! The foregoing formulae are relevant from some points of view. For example, (4.10) and (4.11) have the advantage that the concentration is taken into account and this is useful when studying optimal control problems and defining optimal therapeutics. Another interesting property comes from:

$$x_1(t,D) = \alpha_1 \exp(\lambda_1 t) + \alpha_2 \exp(\lambda_2 t) + \alpha_3 \exp(\lambda_3 t)$$

where α_1, α_2, α_3, λ_1, λ_2, λ_3 are identified as functions of D.

Substituting this in (4.10) gives:

$$x_2(t,D) = b_0 + b_1(D)\log(1 + x_1(t,D))$$
$$+ b_2(D)\log^2(1 + x_1(t,D))$$

(4.20)

and it is clear that x_2 may be obtained for an arbitrary D at any time t without a new experimental measure.

Let us now examine (4.18). The effect $x_2(t)$ is a function of only t and D. Therefore, knowing only the dose D and the time t, the effect can be calculated without any knowledge of $x_1(t)$, the measurement of which can thus be avoided. But later we shall see that the concentration plays an important role in optimal control problems associated with real therapeutics. In fact, when several doses are administered, the concentrations may be additive.

We come back to the search for a simple and appropriate relationship. When considering multiple absorption one must be able to write the resultant effect. The most natural hypothesis [72] is to add the concentrations associated with each absorption. Because of this, and because the purpose of this study is to obtain an optimization which takes account of repeated administration, the previously obtained formula has to be modified as follows:

$$x_2(t) = b_0 + b_1 \log(1 + x_1(\gamma_D(t)))\qquad(4.21)$$

with b_1 a constant (independent of D) and $\gamma_D(t)$ an unknown function of t and D which has to be identified from experimental data. Then $x_2(t)$ depends on D due to the function $\gamma_D(t)$ which necessarily satisfies $\gamma_D(0) = 0$. γ_D plays the part of a time lag and allows for the fact that an effect can remain even though no trace of the drug can be found. This fact can be established by considering the previous table (4.B).

There are many possibilities for finding the function $\gamma_D(t)$. For example, it is easy to use:

(i)   a linear function:
$$\gamma_D(t) = A(D) \cdot t$$
        where the coefficient A has to be identified as a function of D.

(ii)   a second degree polynomial:
$$\gamma_D(t) = A_1(D)t + A_2(D)t^2$$
        where A_1 and A_2 are functions of D to be found from experimental data

(iii) a third degree polynomial:
$$\gamma_D(t) = A_1(D)t + A_2(D)t^2 + A_3(D)t^3$$
        with $A_i(D)$ to be identified.

The best approximation is of course given by the third degree polynomial, although acceptable approximations are given by both the other functions. The function $x_1(t)$ in formula (4.21) is:

$$x_1(t) = a_1 \exp(\lambda_1 t) + a_2 \exp(\lambda_2 t)$$

where the a_1, a_2, λ_1, λ_2 are calculated from the data corresponding to D = 2 mg. The dependence of x_2 on D is carried through $\gamma_D(t)$. Thus (4.21) may be used to define optimal therapeutics but the corresponding mathematical problem has still to be explained.

4.4 Optimal therapeutics

To define optimal therapeutics is is necessary to be precise
about the criterion chosen. There are two types of
criteria: continuous and discrete. Let us begin with the
continuous type:

$$J = \int_0^T (x_2(t) - A)^2 dt \qquad (4.22)$$

Where A is the desired value of the effect x_2. We want to
minimise J as a function of the doses D_i given at times t_i.
In other words we have to solve the following problem:

$$\underset{D_0,\ldots,D_m,\ t_0,\ldots,t_m}{\text{Min}} \int_0^T (x_2(t) - A)^2 dt \qquad (4.23)$$

where D_i is given at t_i, $t_0 = 0$, $t_m \leqslant T$.

Because of the multiple doses given at the t_i, the
function $x_2(t)$ in (4.23) must be defined. The first
reasonable hypothesis [72] involves the addition of
concentrations in the course of time. In other words, we
have $x_2^i(t)$ in (t_i, t_{i+1}) given by:

$$x_2^i(t) = b_0 + b_1 Log(1 + \sum_{k=0}^i x_1(\gamma(D_k, t - kh))) \qquad (4.24)$$

where h is the interval between two consecutive t_i, that is
$t_{i+1} - t_i = h$. To be more precise, $x_1(\gamma(D_k, t - kh))$ would
be replaced by $Y(t - kh)x_1(\gamma(D_k, t - kh))$ because $x_1(\ . \ )$
due to D_k is equal to zero for $t \leqslant kh$.

Another method is to add the effects. Then the
following relation is satisfied:

$$x_2^i(t) = b_0 + \sum_{k=0}^i b_1 Log(1 + x_1(\gamma(D_k, t - kh))) \qquad (4.25)$$

The formula (4.24) or (4.25) may often be simplified because
the concentrations or effects generally disappear after 24
or 28 hours. Therefore in (4.24) or (4.25) the sum includes
only two or three terms at most (in fact, the last two or

three terms). In practice A is chosen to be 19 or 20 for a beta-blocker such as Tertatolol (Dr. Prost, Labo Servier). h is taken as 12 or 24 hours. There are several numerical methods for solving the problem of (4.23). We shall describe them successively.

a) A simplified technique

This consists of fixing the times t_i (with $t_{i+1} - t_i =$ const.) and solving the following optimization problems:

$$\underset{D_j}{\text{Min}} \int_{t_j}^{t_{j+1}} (x_2^j(t) - A)^2 dt, \quad j = 0, \ldots, m \qquad (4.26)$$

It is a simple minimization problem in terms of the single variable D_j. It is clear that the solution of (4.26) gives only a sub-optimum. We do not have the global optimum, but generally only a local, though interesting, one.

b) Utilization of dynamic programming technique [5]

Let us assume that times of absorption are fixed (equidistant). Then (4.23) becomes:

$$\underset{D_0, \ldots, D_m}{\text{Min}} \int_0^T (x_2(t) - A)^2 dt$$

$$= \underset{D_0, \ldots, D_{m-1}}{\text{Min}} \ [\int_0^{t_1} (x_2^0(t) - A)^2 dt + \ldots +$$

$$+ \underset{D_m}{\text{Min}} \int_{t_m}^{t_{m+1}} (x_2^m(t) - A)^2 dt] \qquad (4.27)$$

This equality is valid because only $\int_{t_m}^{t_{m+1}}$ depends on D_m and we have the relation:

$$\underset{D_0, \ldots, D_m}{\text{Min}} = \underset{D_0, \ldots, D_{m-1}}{\text{Min}} \ (\ \underset{D_m}{\text{Min}} \ )$$

for all functions.

Developing the decomposition gives finally:

$$\underset{D_0,\ldots,D_m}{\text{Min}} \int_0^T (x_2(t) - A)^2 dt = \underset{D_0}{\text{Min}} \left[\int_{t_0}^{t_1} (x_2^0(t) - A)^2 dt \right.$$

$$+ \underset{D_1}{\text{Min}} \left[\int_{t_1}^{t_2} (x_2^1(t) - A)^2 dt + \ldots + \right.$$

$$+ \underset{D_m}{\text{Min}} \left[\int_{t_m}^{t_{m+1}} (x_2^m(t) - A)^2 dt \right] \ldots]]$$

$$(4.28)$$

A numerical algorithm is as follows:

(i)   At step 0, the functional:

$$\int_{t_m}^{t_{m+1}} (x_2^i(t) - A)^2 dt$$

is minimised as a function of D_m and for different values of D_{m-1} so that the following function:

$$D_j = \varphi_0(D_{j-1}) \qquad (4.29)$$

may be identified for instance by the expression:

$$\varphi_0(D_{j-1}) = A_1 \exp(\lambda_1 D_{j-1}) + A_2 \exp(\lambda_2 D_{j-1})$$

In   (4.29) it is assumed that $x_2^j(t)$ involves only two terms and thus depends only on D_j and D_{j-1}.   This is a consistent simplification for our optimal control problem.

(ii) At step m the sum of the following integrals is minimised in terms of a single variable D_0:

$$\underset{D_0}{\text{Min}} \left[\int_{t_0}^{t_1} (x_2^0(D_0) - A)^2 dt + \int_{t_1}^{t_2} (x_2^1(D_0, \varphi_{m-1}(D_0)) - A)^2 dt + \right.$$

$$\ldots + \int_{t_m}^{t_{m+1}} (x_2^m(\varphi_1(D_{m-2}), \varphi_0(D_{m-1})) - A)^2 dt]$$

$$(4.30)$$

The functions φ_0, φ_1, . . . , φ_{m-1} are identified successively at steps 0, 1, . . ,m-1.

Since D_0 is determined by solving the optimization system (4.30) we can deduce all the other doses. For example:

$$D_1 = \varphi_{m-1}(D_0), \quad D_2 = \varphi_{m-2}(D_1) \quad \text{and so on.}$$

Remark If more than two influential terms in $x_2^i(t)$ are considered, the functions φ_i include several variables and their identification is much more complicated (numerically speaking).

c) Using a global optimization method [15]

Now we want to solve the general problem:

$$\underset{D_0,\ldots,D_m,\; t_0,\ldots,t_m}{\text{Min}} \int_0^T (x_2(t) - A)^2 dt \qquad (4.31)$$

where both D_i and t_i (i = 1, . . ,m) are unknown. As already described, we can reduce the unknowns to a single variable using the properties of the Archimedean spiral. Then we obtain an optimization problem in a single variable θ. To be more precise, we first make the Alienor transformation described below. Then a densification principle is applied. A relative minimum is calculated using a probabilistic technique. Afterwards we advance from $\theta = 0$ and look for successive minima. The lowest obtained is the required approximate optimum. To refine it the local variation method (Chapter 2) is used to obtain a better approximation to the global minimum.

For example if we have 5 variables called for this purpose x_i, i = 1, . . ,5, the transformation formulae are as follows:

$$\theta_0 = \theta\cos\theta/2\pi \quad , \quad \theta_1 = \theta\sin\theta/2\pi \quad , \quad \theta_2 = \theta_0\cos\theta_0/2\pi$$

$$\theta_3 = \theta_0\sin\theta_0/2\pi \quad , \quad \theta_4 = \theta_1\cos\theta_1/2\pi \qquad (4.32)$$

Then:

$$y_1 = \theta_2\cos\theta_2/2\pi \quad , \quad y_2 = \theta_2\sin\theta_2/2\pi,$$

$$y_3 = \theta_3\cos\theta_3/2\pi \quad , \quad y_4 = \theta_3\cos\theta_3/2\pi \quad , \quad y_5 = \theta_4\cos\theta_4/2\pi$$

are the transformed variables. The densification principle
is as follows. We make m_i and M_i the lower and upper bounds
of the variables x_i. The densification principle is such
that:

$$\text{if } 0 < y_i < 1/2 \quad \text{then } z_i = m_i + (M_i - m_i)2y_i$$

otherwise:

$$z_i = m_i + (M_i - m_i)(2y_i - \text{INT}(2y_i))$$

where $i = 1, \ldots, 5$ which is used for moving in R^5.

d)   A method for fixing the intervals containing the t_i

Firstly, we suppose that the t'_i ($i = 0, \ldots, m$) are given,
and that the t_i (unknown) satisfy $t'_i \leqslant t_i \leqslant t'_{i+1}$. We are
looking for the $t_i \in [t'_i, t'_{i+1}]$ and the doses D_i such that:

$$\int_0^T (x_2(t) - A)^2 dt \qquad (4.33)$$

be minimised.

     This problem may be solved using the following
iterative method:

(i)   The first step is to minimise:

$$\int_{t'_0}^{t'_1} (x_2(t) - A)^2 dt \qquad (4.34)$$

with respect to dose D_0 given at time $t_0 = t'_0 = 0$. The
function $x_2(t)$ is calculated from (4.24) or (4.25).

(ii) The second step is to consider the optimization problem:

$$\underset{D_1}{\text{Min}} \int_{t'_1}^{t'_2} (x_2(t) - A)^2 dt = \underset{D_1}{\text{Min}} \int_{t_1}^{t'_2} (x_2(t) - A)^2 dt + \alpha \qquad (4.35)$$

where α is a constant independent of D_i:

$$\left(\alpha = \int_{t'_1}^{t_1} (x_2(t) - A)^2 dt \right)$$

Calculating the function D_1 for several values of t_1, the function $D_1 = \varphi_1(t_1)$ may be identified numerically.

Then having the function φ_1 we solve the problem:

$$\text{Min}_{t_1} \ [\int_{t_1'}^{t_1} (x_2(t) - A)^2 dt + \int_{t_1'}^{t_2'} (x_2(t) - A)^2 dt] \qquad (4.36)$$

where the values D_0 determined and $\varphi_1(t_1)$ calculated are inserted in the second integral. The problem (4.36) is a minimization problem in terms of the single variable t_1. This process uses the fundamental relation:

$$\text{Min}_{D_1, t_1} F \ = \ \text{Min}_{t_1} \ [\text{Min}_{D_1} F]$$

(iii)   The algorithm is continued to step m at which φ_1, φ_2, . . . , φ_{m-2} have been calculated. First the following problem is solved numerically:

$$\text{Min}_{D_{m-1}} \int_{t_{m-1}}^{t_m'} (x_2(t) - A)^2 dt \qquad (4.37)$$

for several values t_{m-1}.

Then the function $D_{m-1} = \varphi_{m-1}(t_{m-1})$ is identified using, for example, a polynomial or exponential structure. Then the solution of:

$$\text{Min}_{t_{m-1}} \ [\int_{t_{m-1}'}^{t_m} (x_2(t) - A)^2 dt + \int_{t_m}^{t_m'} (x_2(t) - A)^2 dt] \qquad (4.38)$$

is possible using an optimization method (dichotomy, for instance). The values $D_0, D_1, . . . , D_{m-2}$ and the function $D_{m-1} = \varphi_{m-1}(t_{m-1})$ are substituted in the function $x_2(t)$ of (4.38).

4.5 Numerical results

The concentration $x_1(t)$ corresponding to the Tertatolol was chosen as:

$$x_1(t) = a_1 \exp(\lambda_1 t) + a_2 \exp(\lambda_2 t)$$

where $a_1 = -a_2 = 54.437$, $\lambda_1 = -1.9178$, $\lambda_2 = -0.246$.

In formula (4.21) $b_0 = 25$ and $b_1 = -2.6875786$, $V_D(t)$ is a third degree polynomial function whose coefficients are calculated from experimental data.

The dynamic programming technique gives the following optimal doses:

With $m = 2$, $T = 96$ hours, $A = 20$, one finds
$D_0 = 4.69$ mg, $D_1 = 4.526$ mg, $D_2 = 4.516$ mg, $D_3 = 4.674$ mg.

The method described in (d) gives, when $m = 1$, $A = 19$:
$D_0 = 5.895$, $D_1 = 5.687$, $t_1 = 28$.

Finally the global optimization method gives the following:

For $A = 19$, $m = 2$, $T = 72$ h, we obtain
$t_0 = 0$, $t_1 = 38.682$, $t_2 = 51.84$,
$D_0 = 8.07$ mg, $D_1 = 1.0$ mg, $D_2 = 5.3$ mg.

For $A = 20$, $m = 2$, $T = 72$ h, we obtain
$t_0 = 0$, $t_1 = 43.56$, $t_2 = 58.89$,
$D_0 = 6.89$ mg, $D_1 = 1.0$ mg, $D_2 = 1.0$ mg.

The practical consequences are as follows:

(i)   A   dose of about 5 mg per 24 hours is almost optimal. This has been confirmed by experience with patients.

(ii) Small changes in the speed of absorption (± 4 h) of the drug have little influence on the effect.

Modelling and optimization methods allow the proposal of more rational therapeutics. It is possible, using these techniques, to minimise the quantity and maximise the effectiveness of drugs.

4.6 Non-linear compartment approach

Formula (4.24) or (4.25) allows the calculation of the constant B corresponding to $x_1(t)$ and associated with $x_2 \equiv A$. We propose to examine the optimal therapeutics using only $x_1(t)$. The new problem is as follows:

$$\underset{D_0,\dots,D_m,\ t_0,\dots,t_m}{\text{Min}} \int_0^T (x_1(t) - B)^2 dt \qquad (4.39)$$

with $t_0 = 0$, $t_m \leqslant T$.

Further, suppose that we have experimental data $x_1(t)$ corresponding to $D_1, \ldots, D_p$. Then we shall try an approximation for $x_1(t)$ as follows:

$$x_1(t) = \sum_{i=1}^{n} a_i(D)\exp(\lambda_i(D).t) \qquad (4.40)$$

where the functions of D, a_i and λ_i must be identified from experimental data.

Admitting the additivity of concentrations with repeated doses D_i at time t_i ($i = 1, \ldots, m$) we obtain:

$$x_1^j(t) = \sum_{k=1}^{j} [\sum_{i=1}^{n} a_i(D_k)\exp(\lambda_i(D_k)(t - t_k))] \qquad (4.41)$$

on (t_j, t_{j+1}).

Optimising therapeutics leads to the problem:

$$\underset{D_0,\ldots,D_m,t_0,\ldots,t_m}{\text{Min}} \sum_{k=0}^{m-1} \int_{t_k}^{t_{k+1}} (x_2(t) - B)^2 dt \qquad (4.42)$$

It is a non-linear optimization problem needing techniques as developed earlier. The numerical results are not very different from those obtained by other techniques (see Section 4.5).

<u>Remark</u> The coefficients $a_i(D)$ and $\lambda_i(D)$ may be determined from the compartmental model illustrated in Figure 12:

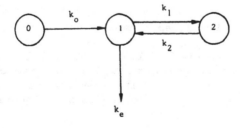

FIG. 12

The experimental data is $x_1(t)$ (a mean value obtained with 10 subjects and four doses, D = 1, 2, 5 and 10 mg). The corresponding equations are:

$$C_0 = (D/V)\exp(-k_0 t)$$

$$C_1 = A_0\exp(-\lambda_0 t) + A_1\exp(-\lambda_1 t) + A_2\exp(-\lambda_2 t)$$

$$C_2 = B_0\exp(-\lambda_0 t) + B_1\exp(-\lambda_1 t) + B_2\exp(-\lambda_2 t)$$

(4.43)

$$A_0 + A_1 + A_2 = B_0 + B_1 + B_2 = 0, \quad V = \text{vol. of comp. 1.}$$

when looking for an explicit solution.

Setting $k_0 = \lambda_2$ we can determine the coefficients of (4.43):

$$B_i = k_1 A_1/(k_2 - \lambda_i)$$

$$V = \lambda_2 D/[A_0(\lambda_2 - \lambda_0) + A_1(\lambda_2 - \lambda_1)]$$

$$= -\lambda_2 D/(A_0\lambda_0 + A_1\lambda_1 + A_2\lambda_2)$$

$$k_2 = [A_0\lambda_1(\lambda_2 - \lambda_0) + A_1\lambda_0(\lambda_2 - \lambda_1)]/[A_0(\lambda_2 - \lambda_0) + A_1(\lambda_2 - \lambda_1)]$$

$$= [A_0\lambda_1\lambda_2 + A_1\lambda_0\lambda_2 + A_2\lambda_0\lambda_1]/[A_0\lambda_0 + A_1\lambda_1 + A_2\lambda_2]$$

$$k_e = \lambda_0\lambda_1/k_2, \quad k_1 = \lambda_0 + \lambda_1 - k_e - k_2, \quad k_0 = \lambda_2$$

(4.44)

A global optimization method allows the calculation of a_i, l_i in:

$$C_1 = a_0\exp(-l_0 t) + a_1\exp(-l_1 t) + a_2\exp(-l_2 t) \qquad (4.45)$$

where $a_2 = -a_0 - a_1$, and $l_0 < l_1 < l_2$

k_0 is chosen equal to the greatest λ_i if k_1 and $k_2 > 0$, or to the intermediate value otherwise. Table 4.C shows some numerical results.

The numerical identification of $a_1(D)$ and $\lambda(D)$ becomes easy.

D	λ_0	A_0	λ_1	A_1	$\lambda_2 = k_0$	A_2	V	k_1	k_2	k_e	error %
1	0.5925	990.000	1.1875	- 0.604	0.6245	-989.40	0.020	0.003148	1.1812	0.596	6.300
2	0.2731	65.122	5.1986	-19.312	1.0497	- 45.81	0.0161	2.6412	2.1791	0.6514	4.238
5	0.0187	4.740	0.3432	314.41	0.8528	-319.15	0.026	0.09331	0.02651	0.2420	6.994
10	0.1375	92.221	0.4391	380.202	1.3064	-472.42	0.0299	0.07975	0.21177	0.2850	3.739

Table 4.C

4.7 Optimal therapeutics using a linear approach

Some authors [65] have considered very simplified problems
such as the following. Suppose a drug is absorbed by a
patient in accordance with a linear compartmental model.
From Chapters 2 and 3 we know that the concentration in the
blood associated with the dose D is given by the relation:

$$x_1(t) = Dx_1^*(t) \qquad (4.46)$$

where $x_1^*(t)$ is the concentration in the blood (compartment
1) corresponding to a dose of 1 mg (x_1^* is assumed to be
measured). Now suppose we are looking for the optimal dose
in 24 hours such that $x_1(t)$ is as near as possible to a
fixed constant a. The mathematical problem is:

$$\underset{D}{\text{Min}} \int_0^{24} (Dx_1^*(t) - a)^2 dt = \underset{D}{\text{Min}} J \qquad (4.47)$$

The optimum D* is obtained by writing a necessary
condition for optimality:

$$\partial J/\partial D = 0 \iff 2D^* \int_0^{24} x_1^{*2}(t)dt - 2a \int_0^{24} x_1^*(t)dt = 0 \qquad (4.48)$$

giving:

$$D^* = a \int_0^{24} x_1^*(t)dt \bigg/ \int_0^{24} x_1^{*2}(t)dt \qquad (4.49)$$

Hence an explicit value is obtained for the optimal dose D*.
A generalization to multiple doses D_i at time t_i is
possible, as follows. To solve the optimal control problem:

$$\underset{D_0,\dots,D_m}{\text{Min}} \int_0^T (x_1(t) - a)^2 dt = \underset{D_0,\dots,D_m}{\text{Min}} \sum_{i=0}^{m-1} \int_{t_i}^{t_{i+1}} (x_1(t) - a)^2 dt$$

$$(4.50)$$

where the t_i are fixed ($t_0 = 0, t_m \leqslant T$), the following
algorithm may be used. Note that it will give only a local
minimum.

(i) First we calculate D_0^* as a solution of:

$$\text{Min} \atop D_0 \quad \int_{t_0}^{t_1} (x_1(t) - a)^2 dt$$

In fact, we have $D_0^* = a \int_{t_0}^{t_1} x_1^*(t) dt / \int_{t_0}^{t_1} x_1^{*2}(t) dt$

(ii) At the second step the criterion:

$$\int_{t_0}^{t_1} (x_1(t) - a)^2 dt + \int_{t_1}^{t_2} (x_1^1(t) - a)^2 dt$$

has to be minimised in terms of D_1. This is equivalent to the minimization of:

$$\int_{t_1}^{t_2} (x_1^1(t) - a)^2 dt$$

in terms of D_1 because the first term does not depend on D_1. x_1^1 has to be evaluated; it is the resultant concentration of D_0 and D_1, that is:

$$x_1^1(t) = x_1^*(t) \cdot D_0^* + D_1 Y(t - t_1) \cdot x_1^*(t - t_1) \qquad (4.51)$$

where $Y(t)$ is the Heaviside function.

As before, the optimisation problem:

$$\text{Min} \atop D_1 \quad \int_{t_1}^{t_2} (x_1^1(t) - a)^2 dt \qquad (4.52)$$

can be solved exactly and we find a D_1^* as quotient of two integrals.

(iii) At step k of iteration, the problem:

$$\text{Min} \atop D_k \quad \int_{t_k}^{t_{k+1}} (x_1^k(t) - a)^2 dt \qquad (4.53)$$

is solved with:

$$x_1^{(k)}(t) = x_1^*(t) D_0^* + D_1^* Y(t-t_1) x_1^*(t-t_1) + D_2^* Y(t-t_2) x_1^*(t-t_2) +$$

$$+ \ldots + D_{k-1}^* Y(t-t_{k-1}) x_1^*(t-t_{k-1}) + D_k^* Y(t-t_k) x_1^*(t-t_k)$$
$$(4.54)$$

The D_k^* is still explicitly calculated by $\partial J/\partial D_k = 0$.

<u>Remark</u> Suppose that in a compartmental system with two compartments (injection) or three compartments (absorption) we measure the concentration $C_1^*(t)$ corresponding to an injection α in compartment 1, and the concentration $C_1(t)$ which corresponds to an oral dose D in compartment 0 (see Figure 12). Since these concentrations are associated with the same drug the following relation is satisfied:

$$C_1(t) = C_1^*/\alpha * k_0(D/V)\exp(-k_0t) \qquad (4.55)$$

This is due to the fact that the u(t) corresponding to an oral dose is equal to $k_0(D/V)\exp(-k_0t)$.

When C_1 and C_1^* have been measured, therefore, the coefficient k_0/V, and k_0 of $k_0(D/V)\exp(-k_0t)$ can be uniquely identified. It is sufficient to solve the convolution equation (4.55). This technique allows us to resolve the uniqueness problem when considering a three-compartment system.

4.8 Optimal control in a compartmental model with time lag

A product such as Tertatolol has a very long-lasting action (even after 24 hours) and classical linear compartmental models are not able to account for this property. It may be useful to introduce time-lag models to explain this persistent effect. Consider the four-compartment model shown in Figure 13.

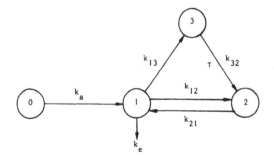

FIG. 13

The corresponding equations for concentrations $C_i(t)$ are:

$$\dot{C}_0 = -k_a C_0$$

$$\dot{C}_1 = k_a C_0 - (k_{13} + k_e + k_{12})C_1 + k_{21}C_2$$

$$\dot{C}_2 = k_{12}C_1 - k_{21}C_2 + k_{31}C_3(t - \tau) \tag{4.56}$$

$$\dot{C}_3 = k_{13}C_1 - k_{31}C_3(t - \tau)$$

where $C_i = x_i/V_1$, x_i = quantity in compartment i, V_1 = volume of compartment 1. The initial conditions are:

$$C_0(0) = D/V_1, \quad C_1(0) = C_2(0) = C_3(0) = 0 \tag{4.57}$$

Furthermore:

$$C_3(t) = 0 \quad \text{for} \quad -\tau \leqslant t \leqslant 0 \tag{4.58}$$

We will now give some further particulars of the model of Figure 13. Three compartments, 0, 1 and 2, are classical and correspond respectively to the absorption, blood, and action compartments. Compartment 3 (see Figure 13) stores the drug. After a time lag τ the drug is transferred towards the second compartment. This can explain the duration of the effect.

(4.56) can be solved by well-tried methods [24]. For instance, the time $(0,\infty)$ may be decomposed into sub-intervals of length τ, that is to say $[0,\tau]$, $[\tau,2\tau]$, · · , $[n\tau,(n+1)\tau]$. The differential system (4.56), (4.57), (4.58) can be solved by classical techniques on each interval beginning with the first (Euler or Runge-Kutta method). In each interval we have to solve a differential system whose terms $C_3(t - \tau)$ have been previously fixed (for the first interval) or calculated at the previous step (for the other intervals).

Then the identification problem arises. It consists of identifying the k_{ij} (exchange parameters), the time lag τ and the volume of compartment 1, i.e. V_1. The experimental data is $C_1(t_j)$ for $j = 1, \ldots, m$. There are more relations than in the linear compartmental case. In fact $C_1(t)$, $C_2(t)$ cannot be equal to finite linear combinations of exponentials $(\exp(\lambda_i t))$. Theoretical identification is therefore more likely to be unique. On $[n\tau,(n + 1)\tau]$ we can use our previous techniques (see Chapter 2). For instance, the Laplace transform may be used, noting that $C_3(t - \tau)$ is equal to:

$$C_3(t - \tau) = \delta_{(t-\tau)} * C_3(t) \qquad (4.59)$$

The Laplace transformation of (4.59) uses:

$$\overbrace{C_3(t - \tau)} = \exp(-\tau t) * \hat{C}_3 \qquad (4.60)$$

Thus relationships may be found between known functions and parameters. They are not rational fractions as for linear systems but are combinations of rational fractions and exponential functions. This is due to terms such as (4.60). Furthermore the Laplace transform applied to (4.56) allows convolution relations to be found between C_0 and C_1, or C_0 and C_2, or C_1 and C_2:

$$C_1 = K_1 * (D/V_1)\exp(-k_a t)$$

$$C_2 = K_2 * (D/V_1)\exp(-k_a t) \qquad (4.61)$$

$$C_2 = K_3 * C_1$$

where the functions K_i are explicitly calculated from (4.56).

Indeed, considering the second relation (4.56) gives directly:

$$C_2 = C_1'/k_{21} + (k_{13} + k_{12} + k_e)C_1/k_{21} \\ - (k_a/k_{21})\exp(-k_a t)(D/V_1)$$

which is a particular type of convolution relation.

The coefficients k_{ij} and τ may be obtained numerically by using the global optimization technique called Alienor (Chapter 3). We relate our unknowns k_{ij}, τ so as to obtain one single variable θ. Since $C_1(t)$ is given at t_j, $j = 1, \ldots, m$, the following functional J is introduced:

$$J = \sum_{j=1}^{m} (C_1(t_j) - C_1^e(t_j))^2 \qquad (4.62)$$

where $C_1(t_j)$ represents the concentration in compartment 1 at time t_j, coming from the solution of (4.56) where θ, and therefore the k_{ij} and τ, are given. C_1^e is the measured concentration. J is a function of the single variable θ and

must be minimised in terms of this variable.   Of course,
$C_1(t_j)$ has to refer to several intervals $[i\tau, (i+1)\tau]$, and on
each of these intervals (4.56) has to be solved.

When the identification is realised, the optimal
control problem associated with (4.56) may be considered.
Several formulations are possible.   They differ in the
choice of criterion to be minimised.

For instance a possible choice is the following:

Find the doses D_i given at unknown times t_i such that:

$$\int_0^T (C_1(t) - A)^2 dt \qquad (4.63)$$

is minimised with A a fixed constant.

Reintroducing the value of $C_1(t)$ obtained in (4.61)
leads to the problem:

$$\underset{D_0,\ldots,D_m,t_0,\ldots,t_m}{\text{Min}} \int_0^T ([K_1 * (D/V_1)\exp(-k_a t)]^* - A)^2 dt \qquad (4.64)$$

where the function [ ]* has to be specified.   Because of
the additivity of concentrations we have:

$$[\ ]^* = K_1 * \sum_{i=0}^m (D_i/V_1).Y(t - t_i).\exp(-k_a(t-t_i)) \qquad (4.65)$$

which is an explicit function of the D_i, t_i.

An optimization method may therefore be applied to find
optimal values for the D_i and t_i.

If another criterion is chosen the same type of problem
arises.   For example:

$$\underset{D_0,\ldots,D_m,t_0,\ldots,t_m}{\text{Min}} \int_0^T (C_2(t) - B)^2 dt, \quad B = \text{constant} \qquad (4.66)$$

is transformed into:

$$\underset{D_0,\ldots,D_m,t_0,\ldots,t_m}{\text{Min}} \int_0^T (K_2 * (D/V)\exp(-k_a t) - B)^{*2} dt$$

where an expression similar to (4.65) is valid for ( )*.

As in the previous approaches the control problem is reduced to a classical optimization system with a functional which explicitly includes the control variables. This transformation is very important to enable us to obtain the optimal solution easily.

CHAPTER 5

GENERAL MODELLING IN MEDICINE

5.1 The problem and the corresponding model

In medicine, we are often confronted with biochemical systems where one or more substances play antagonistic roles. For example, this is the case when studying hormonal systems which are fundamental in human and animal physiology. They maintain many physiological regulations and a malfunction is catastrophic for an organism. In this chapter, we shall present a system first developed by B. Weil [74], and completed by A. Guillez [58] and P. Nelson [53].

The first question asked by B. Weil was: how is it possible to treat an hormonal imbalance in cancerous subjects? In fact the following observation was made by many physicians when treating patients: administration of the missing hormone is not sufficient to improve the condition of the patient because of the existence of some regulatory mechanisms in the human system. The biological system with all its generality and complexity has to be taken into account.

In the following we shall describe a particular system (the endocrine system) which plays an important role in the development of cancer. In contrast to the usual practice, two endocrinal axes will be taken into account. In the classical literature each endocrinal sub-system is modelled separately, and no connection is established between the two sub-systems. Therefore, therapeutics cannot be defined mathematically and so the failure of many cancer treatments can be explained. Common sense is often not sufficient for understanding complex systems nor for defining appropriate therapeutics. Modelling and optimal control techniques are necessary for the definition of optimal biochemical treatments.

Coming back to our concrete problem we want to propose mathematical equations describing the interaction between two hormones: cortisone called $x(t)$ and vasopressin called $y(t)$. In fact we want to relate the concentrations $x(t)$, y(t) and the control variables X(t) and Y(t). X(t) and Y(t) are injected from the outside and are often called exogenous variables. They are of the same type as x(t) and y(t) respectively. The final objective may be described by considering the following experimental observation: the normal (non-cancerous) subject has an hormonal balance. That is:

$$x(t) = y(t)$$

$$x + y - m = 0$$

(5.1)

where m is a constant (first approximation) depending on the subject.

Since a balance is observed in the normal subject, what happens during the cancerous condition? In fact, a lack of balance comes about:

$$x(t) \neq y(t)$$

$$x + y - m \neq 0$$

(5.2)

Therefore an idea for therapy arises. It is to redress the balance by means of injections of hormones. The associated optimal control problem will be described later.

After many tests a non-linear differential system was chosen:

$$\dot{x} = k_1 u + k_2 u^2 + c_1 v + c_2 v^2$$

$$\dot{y} = k_3 u + k_4 u^2 + c_3 v + c_4 v^2$$

(5.3)

with initial conditions (known or unknown in individual cases) and where $x - y = u$, $x + y - m = v$, and where x(t) = cortisone and y(t) = vasopressin. A. Guillez has suggested introducing third degree terms to facilitate the study of asymptotic stability [58]. This is not very relevant to the examination of optimal therapeutic stability and will not be discussed here.

5.2 The identification problem

First notice that (5.3) is a simulation model (see Chapter 1) since the coefficients k_i, c_i do not have biological reality. These parameters have to be identified from experimental data. Two cases are possible:

a) $x(t)$ and $y(t)$ are measured for $t = t_j$, $j = 1, \ldots ,M$. Then approximations may be made. For example, $x(t)$ and $y(t)$ can be approximated by a Fourier series [60] or by an exponential approximation. To demonstrate the identification method let us choose:

$$x(t) = \sum_{i=1}^{n} a_i \exp(\lambda_i t)$$

$$(5.4)$$

$$y(t) = \sum_{i=1}^{n} b_i \exp(\mu_i t)$$

where the a_i, λ_i, b_i, μ_i must be identified from experimental data using an optimization technique.

Substituting (5.4) into (5.3) gives the following functional system:

$$a_i \lambda_i \exp(\lambda_i t) = k_1 \left(\sum_i a_i \exp(\lambda_i t) - \sum_i b_i \exp(\mu_i t) \right)$$

$$+ k_2 \left(\sum_i a_i \exp(\lambda_i t) - \sum_i b_i \exp(\mu_i t) \right)^2$$

$$+ c_1 \left(\sum_i a_i \exp(\lambda_i t) + \sum_i b_i \exp(\mu_i t) - m \right)$$

$$+ c_2 \left(\sum_i a_i \exp(\lambda_i t) + \sum_i b_i \exp(\mu_i t) - m \right)^2$$

$$b_i \mu_i \exp(\mu_i t) = k_3 \left(\sum_i a_i \exp(\lambda_i t) - \sum_i b_i \exp(\mu_i t) \right)$$

$$+ k_4 \left(\sum_i a_i \exp(\lambda_i t) - \sum_i b_i \exp(\mu_i t) \right)^2$$

$$+ c_3 \left(\sum_i a_i \exp(\lambda_i t) + \sum_i b_i \exp(\mu_i t) - m \right)$$

$$+ c_4 \left(\sum_i a_i \exp(\lambda_i t) + \sum_i b_i \exp(\mu_i t) - m \right)^2$$

$$(5.5)$$

Since we know a_i, b_i, λ_i, μ_i the system (5.5) may be written for $t = t_j$, $j = 1, \ldots ,N$. Note that the parameters will be identified from measurements made at times $t = t_j$, $j = 1, \ldots ,M$. When (5.5) is used, the points t_j may be chosen arbitrarily. The a_i, b_i, λ_i, μ_i are obtained by minimising the two functionals:

$$J_1 = \sum_{j=1}^{M} [x(t_j) - \sum_{i=1}^{n} a_i \exp(\lambda_i t_j)]^2$$

$$\text{and} \quad J_2 = \sum_{j=1}^{M} [y(t_j) - \sum_{i=1}^{n} b_i \exp(\mu_i t_j)]^2$$

The writing of (5.5) for $t = t_j'$, $j = 1, \ldots, N$ provides $2N$ algebraic linear functions of the parameters k_i, c_i, $i = 1, \ldots, 4$. Let us write these as follows:

$$f_i(k_1, k_2, k_3, k_4, c_1, c_2, c_3, c_4) = f_i(k_j, c_j) = 0$$

$$i = 1, \ldots, 2N \tag{5.6}$$

This redundant linear system may be solved by using the functional:

$$J = \sum_{i=1}^{2N} f_i^2(k_j, c_j) \tag{5.7}$$

which is a second degree polynomial.

Solving (5.6) is equivalent to looking for the minimum of J as a function of the k_i, c_i and satisfying $J = 0$. Numerically it is sufficient to obtain $J = \epsilon$ where $\epsilon > 0$ is a small number chosen by the user of the method.

With this particular function J an explicit method of solution may be provided. The relations:

$$\partial J / \partial k_i = \partial J / \partial c_i = 0 , \quad i = 1, \ldots, 4 \tag{5.8}$$

furnish eight linear algebraic relations whose solution give the desired optimum. Now let us examine the second possibility.

b)   If only $x(t)$ (or $y(t)$) is measured, the foregoing development is not applicable. However (5.4) is still valid with only the a_i and λ_i identified using an optimization method. The parameters b_i, and μ_i remain unknown. The relations (5.5) are also true but b_i and μ_i are extra unknowns. We still have relations (5.6) written as follows:

$$f_i(k_j, c_j, b_j, \mu_j) = 0, \quad i = 1, \ldots, 2N \tag{5.9}$$

which are strongly non-linear. If the associated functional:

$$J_1 = \sum_{i=1}^{2N} f_i^2(k_j, c_j, b_j, \mu_j) \qquad (5.10)$$

is minimised as a function of the k_j, c_j, b_j, μ_j it gives the required solution if $J_1 = 0$ at the optimum. The minimization of J_1 requires the use of a local or global optimization technique. It may be verified that the relations $\partial J_1/\partial k_j = \ .\ \ .\ \ . = \partial J_1/\partial \mu_j = 0$ do not provide more simple relations than (5.9).

This numerical method may show the existence of a solution to our identification problem. As the reader can easily verify uniqueness is never ensured. To ensure uniqueness it is necessary to use an energy criterion. With this hypothesis the problem becomes:

Find k_j, c_j, b_k, μ_k , $j = 1, \ .\ .\ .\ ,4$, $k = 1, \ .\ .\ .\ ,N$ such that:

$$\int_0^\infty y^2(t)dt \text{ is minimised under constraints (5.9) that is}$$

$$f_i(k_j, c_j, b_k, \mu_k) = 0, \qquad i = 1,\ .\ .\ ,2N \qquad (5.11)$$

Solution of problems such as (5.11) is possible using:

(i) Penalty functions [18], [46]. To solve (5.11) the following functional is introduced:

$$J_\epsilon = \int_0^\infty y^2(t)dt + \sum_{i=1}^{2N} (1/\epsilon_i) f_i^2(k_j, c_j, b_k, \mu_k) \qquad (5.12)$$

which has be minimised in terms of the unknowns, where

$\epsilon = (\epsilon_1, \ .\ .\ .\ ,\epsilon_{2N})$ is a fixed vector of R^{2N}.

The function J_ϵ is everywhere differentiable provided the f_i and $\int_0^\infty y^2(t)dt$ possess this property. For $\int_0^\infty y^2(t)dt$ and also for the f_i this is obviously the case. Thus problem (5.11) involving constraints is replaced by an unconstrained optimization problem :

$$\underset{k_j, c_j, b_k, \mu_k}{\text{Min}} \quad J_\epsilon \qquad (5.13)$$

The second term in (5.12) is the penalty function. The solution of (5.13) depends on ϵ and if we denote by:

$$x_\epsilon^* = (k_j^*, \ c_j^*, \ b_k^*, \ \mu_k^*)$$

the point giving the minimum of J_ϵ (Min $J_\epsilon(x) = J_\epsilon(x_\epsilon^*)$) then a convergence result can be proved. In fact, when $\epsilon \rightarrow 0$, $x_\epsilon^* \rightarrow x^*$ (x^* as defined above) and is the solution of problem (5.11).

(ii) Uzawa technique [18], [46]. This comes back to the search for a saddle point. The following functional is introduced:

$$J_1(\lambda, x) = \int_0^\infty y^2(t)dt + \sum_{i=1}^{2N} \lambda_i f_i \qquad (5.14)$$

and the optimum of J corresponds to a saddle point (λ^*, x^*) of the functional (5.14). We recall that a saddle point of $J_1(\lambda, x)$ satisfies the relations:

$$J_1(\lambda, x^*) \leqslant J_1(\lambda^*, x^*) \leqslant J_1(\lambda^*, x) \qquad (5.15)$$

for all λ and x.

For further details about this technique it is necessary to consult the specialist literature [46]. We should note that in practice, finding the saddle point of J_1 requires the use of an optimization technique [18], such as, for example, the gradient technique or the Alienor method.

Remark Identification of the unknown parameters of (5.3) may be done using more classical techniques. One of them is to minimise:

$$J_2 = \sum_{j=1}^m (x(t_j) - x^c(t_j))^2 + \sum_{j=1}^m (y(t_j) - y^c(t_j))^2 \qquad (5.16)$$

with respect to the k_i, c_i, where $x(t)$, $y(t)$ are measured at t_j, $j = 1, \ldots, m$. In the functional J_2, $x^c(t)$ and $y^c(t)$ are the calculated concentrations obtained by solving the differential system (5.3) with k_i and c_i fixed. A local optimization method may be used. Similarly, a global optimization technique is suitable. For the latter the (k_i,c_i) have to be transformed into a single variable θ by the Archimedean transformation and J_2 minimized as a function of this variable. When only $x(t)$ is measured the functional J_2 has to be modified by introducing:

$$\int_0^T y_c^2(t)dt \quad \text{instead of the term} \quad \sum (y(t_j) - y^c(t_j))^2.$$

y_c^2 represents the calculated function $y^2(t)$ when the unknown parameters are fixed.

Once the model is identified its optimal control is possible.

5.3 A simple method for defining optimal therapeutics

The method is based on previous ideas of associating a series of linear compartmental models with non-linear situations. First let us suppose that the therapeutics involve injections of doses D_i at time t_i. Suppose also that $x(t)$, $y(t)$ are measured for different doses D_k given at time $t = 0$. Then $x(t)$ and $y(t)$ are approximated by the following:

$$x(t) = a + \sum_{i=1}^n a_i \exp(\lambda_i t)$$

$$(5.17)$$

$$y(t) = b + \sum_{i=1}^n b_i \exp(\mu_i t)$$

with, generally, $n = 2$, 3 or 4 and with two constants introduced to take account of concentrations (> 0) in the blood at any time, even without injections. The λ_i, μ_i are assumed negative and therefore $x(t) \to a$, $y(t) \to a$ as $t \to \infty$. Further, note that $D_k = (D_{k1}, D_{k2})$ is a vector corresponding to the two hormones.

With the previous data, one can say that $a = b = m/2$ for a normal subject. The unknown coefficients a, a_i, λ_i, b, b_i, μ_i are identified as functions of the dose D from experimental data using an optimization technique.

Generally, a and b are independent of the dose D. Since x(t) and y(t) are hormone concentrations we can assume them to be additive when considering multiple doses D_i at times t_i. The optimal control problem is as follows: .

Find doses D_i and times t_i such that:

$$J = \int_0^T [(x^*(t) - m/2)^2 + (y^*(t) - m/2)]^2 dt \text{ is minimised}$$

where $0 \leqslant t_0 \leqslant t_1 \leqslant t_2 \leqslant \ldots \leqslant t_m \leqslant T$ (5.18)

The functions $x^*(t)$, $y^*(t)$ represent hormone concentrations resulting from multiple doses D_i. By analogy with previous developments we may write:

$$x^*(t) = \sum_{k=0}^{j} [a(D_{k1}) + \sum_{i=1}^{n} a_i(D_{k1})\exp(\lambda_i(D_{k1})(t - t_k))]$$
$$\text{on the interval } (t_j, t_{j+1})$$

(5.19)

$$y^*(t) = \sum_{k=0}^{j} [b(D_{k2}) + \sum_{i=1}^{n} b_i(D_{k2})\exp(\mu_i(D_{k2})(t - t_k))]$$
$$\text{on the same interval.}$$

The additivity of concentrations is expressed in (5.19). Its main interest is the explicit dependence of x^*, y^* on the doses D_i and the times t_i. It follows that a classical optimization technique (local or global) may be used to find the minimum of J given by (5.18). The difficult optimal control problem is transformed into a relatively simple minimization problem. In the following we shall give more sophisticated techniques using the differential system (5.3) directly.

5.4 The Pontryagin method [3], [11]

We present this method in the general formulation of Bolza. It can easily be seen that our present optimization problem is a special case. Consider a general set of n differential equations with m control variables $u_1, \ldots, u_m$:

$$x(t) = f(x_1, \ldots, x_n, u_1, \ldots, u_m, t)$$

(5.20)

with initial conditions $x(t_0) = x_0$.

The optimal control problem is to choose $u(t)$ to minimise:

$$J(u) = g[x(t_1),t_1] + \int_{t_0}^{t_1} F(x,u,t)dt \qquad (5.21)$$

subject to (5.20). The functions g, F are known to be continuous and everywhere differentiable. Parameters t_0 and t_1 are fixed. The Pontryagin principle and technique is based on the use of calculus of variations. First recall the definition of differentiability:

The functional $J(u)$ is differentiable at u if:

$$J(u + \delta u) - J(u) = \delta J(u,\delta u) + \epsilon(u,\delta u) \qquad (5.22)$$

where δJ is linear in δu and $\epsilon(u,\delta u) \to 0$ as $\|\delta u\| \to 0$ (using any suitable norm). δJ is called the variation of J corresponding to the variation δu of u. Furthermore we say that u* is an extremum and J has a relative minimum if the following relation is satisfied:

There exists $\epsilon > 0$ such that:

$$\|u - u*\| \leq \epsilon \text{ implies } J(u) - J(u*) \geq 0 \qquad (5.23)$$

A well-known result states that u* is extremal when $\delta J(u*,\delta u) = 0$. It is a necessary (but generally not sufficient) condition for obtaining an extremum.

We shall apply this result to the functional (5.21) and for this we introduce a Lagrange vector $p = (p_1, \cdot \cdot ,p_n)$. We form the following augmented functional incorporating the constraints:

$$J_1 = g[x(t_1),t_1] + \int_{t_0}^{t_1} [F(x,u,t) + p(f - \dot{x})]dt \quad (5.24)$$

Integrating the last term, by parts, gives:

$$J_1 = g[x(t_1),t_1] - [px]_{t_0}^{t_1} + \int_{t_0}^{t_1} [H + \dot{p}x]dt \qquad (5.25)$$

where the Hamiltonian function H is defined by:

$$H(x,u,t) = F(x,u,t) + p.f \qquad (5.26)$$

If $u(t)$ is supposed differentiable on $t_0 \leqslant t \leqslant t_1$, the variation δJ_1 corresponding to δu is given by:

$$\delta J_1 = [(\partial g/\partial x - p)\delta x]_{t=t_1}$$

$$+ \int_{t_0}^{t_1} [\partial H/\partial x \delta x + \partial H/\partial u \delta u + \dot{p}\delta x]\delta t \qquad (5.27)$$

where δx is the variation in x due to δu. We used the notation:

$$\partial H/\partial x = [\partial H/\partial x_1, \ldots, \partial H/\partial x_n] \qquad (5.28)$$

Since $x(t_0)$ is fixed, $[\delta x]_{t=t_0} = 0$ this term does not appear in (5.27).

To remove the term in (5.27) involving δx we may choose:

$$\dot{p}_i = - \partial H/\partial x_i, \quad i = 1, \ldots, n$$

$$p_i(t_1) = (\partial g/\partial x_i)_{t=t_1} \qquad (5.29)$$

Therefore the equation (5.27) is reduced to:

$$\delta J_1 = \int_{t_0}^{t_1} (\partial H/\partial u . \delta u) dt \qquad (5.30)$$

A necessary condition for u^* to be an extremal of J_1 is that:

$$\delta J_1 = 0 \qquad (5.31)$$

which implies:

$$\partial H/\partial u = 0 \qquad (5.32)$$

because (5.30) is equal to zero whatever the value of δu.

To summarise the situation, we see that u^* is an extremum if (necessary conditions):

$$\dot{p}_i = - \partial H / \partial x_i, \quad i = 1, \ldots, n$$

$$p_i(t_1) = (\partial g / \partial x_i)_{t=t_1} \qquad\qquad (5.33)$$

$$(\partial H / \partial u_i)_{u=u^*} = 0 \quad \text{for } t_0 \leqslant t \leqslant t_1,$$
$$i = 1, \ldots, m$$

The $p_i(t)$ are called adjoint variables and the equations $\dot{p}_i = -\partial H / \partial x_i$ are termed adjoint equations.

In conclusion, the initial control problem (5.20), (5.21) is reduced to the solution of (5.33), (5.30), which is a set of differential and partial differential equations. In general, explicit solutions do not exist, and numerical methods have to be used.

Remark   In real problems the control variables are often subject to constraints such as $|u_i| \leqslant k_i$. The foregoing developments have to be slightly modified.   We first introduce the notion of admissible control, which means that the control satisfies the constraints.   Then we consider variations δu such that $u^* + \delta u$ is admissible and $\| \delta u \|$ is chosen sufficiently small that the sign of:

$$J(u^* + \delta u) - J(u^*)$$

is determined by δJ, the differential in (5.22).

Our previous result, giving a necessary condition for an extremum, becomes:

$$\delta J(u^*, \delta u) \geqslant 0 \qquad\qquad (5.34)$$

The development then proceeds as before.   The only difference is a modification in the expression for J_1.

$$\delta J_1(u, \delta u) = \int_{t_0}^{t_1} [H(x, u+\delta u, p, t) - H(x, u, p, t)] dt \qquad (5.35)$$

Applying the necessary condition (5.34) to the expression (5.35) gives:

$$H(x^*, u^* + \delta u, p^*, t) \geqslant H(x^*, u^*, p^*, t) \qquad (5.36)$$

because (5.35) is valid whatever the value of δu.

Here the Pontryagin principle requires that H is minimum at $u = u^*$.   The conditions (5.33) are true except for the third which is replaced by (5.36). The reader may find the relations (5.33) associated with the hormonal problem (5.3), (5.16).

5.5 A simplified technique giving a sub-optimum

Consider the general optimization problem:

$$\dot{x} = g(x,u,t), \quad t \in [0,T]$$

$$x(T) = a \tag{5.37}$$

$$\text{Minimize} \quad J = \int_0^T f(x,u,t)dt \quad \text{as function of u.}$$

where the functions f and g are given, as well as the constants a, T. Note that in (5.37) it is the final state x(T) which is fixed. This is appropriate for our hormonal system.

The numerical algorithm we propose is based on discretization of the differential system of (5.37).

$$x_n = G(x_{n+1}, u_{n+1}, t_{n+1}), \quad n = 0, \ldots, Nh \tag{5.38}$$

where x_n means x(nh) and t_n = nh (h is the discretization step of the explicit method). N is such that Nh = T, and G is explicitly calculable from the function G.

Since we have:

$$\text{Min } J(x,u) = \text{Min}(\int_0^{(N-1)h} f(x,u,t)dt + \int_{(N-1)h}^{Nh} f(x,u,t)dt) \tag{5.39}$$

and furthermore, u(t) may be approximated by a step function equal to u_n on the interval $[(n-1)h, nh]$, the approximation:

$$\int_{(N-1)h}^{Nh} f(x,u,t)dt \simeq h\, f(x_N, u_N, t_N) \tag{5.40}$$

gives the following relation:

$$\text{Min}_{u} \quad J(x,u) = \quad \text{Min}_{u_1,\ldots,u_N} \quad J(x,u)$$

$$= \text{Min}_{u_1,\ldots,u_N} \left(\int_0^{(N-1)h} f(x,u,t)dt + hf(x_N,u_N,t_N) \right)$$

$$= \text{Min}_{u_N} hf(x_N,u_N,t_N) + \text{Min}_{u_1,\ldots,u_{N-1}} \int_0^{(N-1)h} f(x,u,t)dt \qquad (5.41)$$

where the last equality depends on the assertion:

$$\int_0^{(N-1)h} f(x,u,t)dt \text{ is independent of } u_N.$$

It is easy to find the minumum u_N^* of $f(x_N,u_N,t_N)$ using an optimization method. This technique may be iterated. Indeed we have:

$$\text{Min}_{u_1,\ldots,u_{N-1}} \int_0^{(N-1)h} f(x,u,t)dt$$

$$= \text{Min}_{u_1,\ldots,u_{N-1}} \left(\int_0^{(N-2)h} f(x,u,t)dt + \int_{(N-2)h}^{(N-1)h} f(x,u,t)dt \right)$$

$$\simeq \text{Min}_{u_1,\ldots,u_{N-1}} \left(\int_0^{(N-2)h} f(x,u,t)dt + hf(x_{N-1},u_{N-1},t_{N-1}) \right)$$

$$= \text{Min}_{u_1,\ldots,u_{N-2}} \int_0^{(N-2)h} f(x,u,t)dt + \text{Min}_{u_{N-1}} hf(x_{N-1},u_{N-1},t_{N-1}) \qquad (5.42)$$

In the last expression x_{N-1} is replaced by $G(x_N,u_N^*,t_N)$ which is calculable because u_N^* was previously determined. The values $u_{N-1}, u_{N-2}, \ldots, u_1$ are calculated successively. This algorithm may be considered as a variant of the dynamic programming technique.

5.6 A naive but useful method [74]

Let us derive from (5.3) a differential system including the control variables X, Y which are of the same type as $x(t)$, $y(t)$. We obtain:

$$\dot{x} = k_1(u+r) + k_2(u+r)^2 + c_1(v+s) + c_2(v+s)^2$$
$$\dot{y} = k_3(u+r) + k_4(u+r)^2 + c_3(v+s) + c_4(v+s)^2$$

$$(5.43)$$

with known initial conditions.

In (5.43) we set $r = X(t) - Y(t)$ and $s = X(t) + Y(t)$.

The basic idea, to obtain an optimization involving:

$$u + r = 0, \qquad v + s = 0 \qquad\qquad (5.44)$$

is to add two other differential equations which ensure (5.44) when a critical point is obtained. To be more precise, the two following equations are set up:

$$\dot{X} - \dot{Y} = k_5(x - y + X - Y)$$
$$\dot{X} + \dot{Y} = C_5(x + y + X + Y - m)$$

$$(5.45)$$

with initial conditions fixed.

Then the coefficients k_5 and C_5 are chosen so that a critical point at $t = \infty$ exists for (5.45). Therefore we have:

$$\dot{X} - \dot{Y} = 0$$
$$\dot{X} + \dot{Y} = 0$$

$$(5.46)$$

A practical consequence is that:

$$x - y + X - Y \rightarrow 0$$
$$x + y + X + Y - m \rightarrow 0, \qquad \text{as } t \rightarrow \infty$$

$$(5.47)$$

From a biological point of view, (5.47) means that an artificial balance is obtained. Dr. B. Weil used this technique with success to treat patients with cancer [74].

Numerical experiments give almost the same results when
using the different methods. A practical consequence is
that one must use the two hormones. Treatment with only the
missing hormone is very bad, making the patient worse. Only
a systemic approach and use of optimal control methods
allows an efficient therapy to be devised.

CHAPTER 6

BLOOD GLUCOSE REGULATION

Diabetes is a severe disease. For a long time insulin has
been used to treat patients and, although insulin pumps are
successful, no automatic optimised treatment exists. We do
not have the possibility of measuring blood glucose level
continuously. Optimal control in real time is therefore
impossible. A great deal of experimental data on patients
is available and it is easy to improve the medical
treatment, based on the experience gained by physicians.
Particularly, one hopes to avoid large variations in blood
glucose level. Optimization methods may be used to maintain
a steady level. We shall begin with experiments on dogs
described in [44].

6.1 Identification of parameters in dogs

A major study of the identification problem and the optimal
control problem is given in [44]. First note that blood
glucose concentration increases when glucose is administered
in mammals. This results in an increase in the plasma
insulin concentration. The insulin increases the rate of
removal of glucose from the plasma compartment and the
concentration in the blood returns to its normal value. As
a first approximation, a linear two-compartment model may be
used to model the system "insulin-glucose". In fact, blood
glucose regulation [65] is an extremely complex system and
therefore a scheme is needed for a mathematical and
numerical study. The administration of glucose results in
an immediate precipitous rise in plasma insulin
concentration due to a direct effect on the cells in the
pancreas. The insulin, in turn, accelerates the removal of
glucose from the plasma and the blood sugar returns to a
normal value of 80 to 100 mg. per 100 ml. (about 1 g/l). In
metabolic diseases, there may be a reduction of the number
of beta cells in the pancreas, and sometimes the organism
has no insulin at all. This serious condition will be
treated later with reference to humans. For a normal
organism (human or animal) a minimal model with two
compartments was introduced by Bolie [8]. and studied

137

further by R. Kalaba [44].
    The model equations are as follows:

$$\dot{x}_1 = -ax_1 + b_2$$

$$\dot{x}_2 = -cx_1 - d_2 \qquad (6.1)$$

$$x_1(0) = c_1 \quad , \quad x_2(0) = c_2$$

where x_1 and x_2 are the deviations from the mean of the extracellular concentrations of insulin and glucose, respectively. The coefficients a, b, c, d are positive and have to be identified from experimental data.

$$x_1(t_i)$$
$$\qquad (6.2)$$
$$x_2(t_i) \qquad i = 1, \ . \ . \ ,m$$

In [44] noisy measurements are assumed, and relatively sophisticated methods – Gauss-Newton, quasi-linearization – are proposed. These approaches do not seem to be necessary in this simple situation. In fact, we shall use the techniques developed previously.

(i) Firstly $x_1(t)$ and $x_2(t)$ are approximated using:

$$x_1(t) = \sum_{i=1}^{2} a_i \exp(\lambda_i t)$$

$$\qquad (6.3)$$

$$x_2(t) = \sum_{i=1}^{2} b_i \exp(\lambda_i t)$$

where the a_i, b_i, λ_i are identified by introducing the functional:

$$J = \alpha_1 \sum_{j=1}^{m} [x_1(t_j) - \sum_{i=1}^{2} a_i \exp(\lambda_i t)]^2$$

$$+ \alpha_2 \sum_{j=1}^{m} [x_2(t_j) - \sum_{i=1}^{2} b_i \exp(\lambda_i t)]^2 \qquad (6.4)$$

where the weights may be chosen (for example $\alpha_1 = \alpha_2 = 1$) so as to take into account the respective influence of the terms including x_1 and x_2. Putting (6.3) into (6.1) gives:

$$\sum a_i \lambda_i \exp(\lambda_i t) = - a \sum a_i \exp(\lambda_i t) + b \sum b_i \exp(\lambda_i t)$$

$$\sum b_i \lambda_i \exp(\lambda_i t) = - c \sum a_i \exp(\lambda_i t) - d \sum b_i \exp(\lambda_i t)$$

$$(6.5)$$

where the derivatives have disappeared. We now introduce the new functional:

$$
\begin{aligned}
J_1 = \beta_1 \sum_{j=1}^{N} & [\sum a_i \lambda_i \exp(\lambda_i t_j) + a \sum a_i \exp(\lambda_i t_j) \\
& \quad - b \sum b_i \exp(\lambda_i t_j)]^2 \\
+ \beta_2 \sum_{j=1}^{N} & [\sum b_i \lambda_i \exp(\lambda_i t_j) + c \sum a_i \exp(\lambda_i t_j) \\
& \quad + d \sum b_i \exp(\lambda_i t_j)]^2 \quad (6.6)
\end{aligned}
$$

and minimise it as a function of a, b, c, d. The optimum satisfies the necessary relations:

$$\partial J_1/\partial a = \partial J_1/\partial b = \partial J_1/\partial c = \partial J_1/\partial d = 0 \qquad (6.7)$$

which are the linear algebraic relations including four equations and four unknowns.

Remark Another simple technique is to use (6.5) and to identify the coefficients of $\exp(\lambda_1 t)$ and $\exp(\lambda_2 t)$ in the two equations. This gives:

$$a_1 \lambda_1 = - a a_1 + b b_1$$

$$a_2 \lambda_2 = - a a_2 + b b_2$$

$$b_1 \lambda_1 = - c a_1 - d b_1$$

$$(6.8)$$

$$b_2 \lambda_2 = - c a_2 - d b_2$$

which give a, b, c, d by solving a linear system. The reader may complete the calculation and obtain the explicit relations for a, b, c, d as functions of λ_1, λ_2.

In [44] numerical results are given, corresponding to the data in Table 6.A.

Time (mins)	0	10	20	30	40	50	60
Insulin μg/ml	73	0	-6.7	-5.0	-3.0	-2.5	-2.0
Glucose mg/100ml	32	9	6.5	5.5	5.0	2.5	0

Time (mins)	70	80	90	100	110	120	130
Insulin μg/ml	6.8	-2.2	1.3	-7.5	-7.7	-4.2	1.0
Glucose mg/100ml	1.0	-4.0	-1.5	-1.0	0.5	0.0	1.5

Time (mins)	140	150	160	170	180
Insulin μg/ml	-2.5	-2.0	-4.7	0.5	-0.7
Glucose mg/100ml	0.5	-1.0	-7.0	2.0	-3.0

Table 6.A

The calculated parameters are given by:

a = 0.195,   b = -0.171,   c = 0.0467,   d = 0.0686

6.2 The human case

When considering human patients some new difficulties arise since only five or six measures are available. Also we only know the glucose concentration. Insulin concentration cannot be measured easily. In the following we shall study diabetic patients who produce no insulin. The following differential system may be proposed:

$$\dot{x} = ax + by$$

$$\dot{y} = \qquad dy \qquad\qquad (6.9)$$

$$x(0) = x_0, \quad y(0) = c_0$$

where x and y represent the deviations of glucose and insulin concentrations, respectively. In contrast to the previous model, here we have c = 0, since in this case there is no production of insulin. Insulin is supplied to the organism by injection. Tables 6.B and 6.C give experimental data obtained from a large number of patients.

	8 hr.	10 hr.	12 hr.	16 hr.	19 hr.	22 hr.
No. of patients	357	165	304	181	322	176
Mean	1.876	1.986	2.008	2.611	2.352	2.984
s.e.	0.051	0.076	0.053	0.088	0.062	0.094
Confidence interval	1.7082, 2.0438	1.7360, 2.2360	1.8336, 2.1824	2.3215, 2.9005	2.1480, 2.5560	2.6747, 3.2933

Table 6.B Glucose-level variation over a day

	8 hr.	10 hr.	12 hr.	16 hr.	19 hr.	22 hr.
No. of patients	358	358	358	358	358	358
Mean	1.36	0.22	1.98	0.53	1.95	0.46
s.e.	0.05	0.03	0.04	0.05	0.05	0.05
Confidence interval	1.2405, 1.5245	0.1213, 0.3187	1.8484, 2.1160	0.3655, 0.6945	1.7855, 2.1145	0.2955, 0.6245

Table 6.C Amounts of insulin injected

For 352 patients, the mean of total injected insulin was 26.52, with a standard error of 0.17, and confidence interval [25.9607, 27.0793].

The numerical data in Tables 6.B and 6.C was supplied by Dr. Poirier of Servier Laboratories.

As in the first model, the coefficients a, b, d have to be identified using the available experimental data, that is:

$$x(t_j) \quad , \quad j = 0, \ldots, 5 \qquad (6.10)$$

Remember that the t_j represent times of injections of insulin where the dose C_j is given at time t_j.

(6.9) may be solved using the Laplace transform. An explicit method is also possible. In fact when C_0 (insulin dose) is given at time $t_0 = 0$, the second equation gives:

$$y(t) = C_0 \exp(dt)$$

The first equation (6.9) becomes easy to solve. The equation:

$$\dot{x} = ax + bC_0 \exp(dt) \quad , \quad x(0) = x_0$$

has the solution:

$$x(t) = x_0 \exp(at) + bC_0/(d-a)(\exp(dt) - \exp(at)) \qquad (6.11)$$

Because of the additivity of concentrations the general formula corresponding to doses C_j at time t_j is as follows:

$$x(t) = x_0 \exp(at)$$

$$+ b/(d-a) \sum_{p=0}^{5} C_p Y(t-t_p)(\exp(d(t-t_p)) - \exp(a(t-t_p))) \qquad (6.12)$$

where $Y(t)$ is the Heaviside function defined previously.

Identification is easier when using (6.12). As usual a numerical functional J can be associated:

$$J = \sum_{p=0}^{5} (x(t_p) - x_e(t_p))^2 \qquad (6.13)$$

where $x(t_p)$ is calculated from (6.12) and $x_e(t_p)$ is the experimental value at time t_p. A classical optimization problem is therefore obtained. It may be solved by using a local or a global minimization technique. With the previous data, the following results are found [45]:

$$a = 0.0343, \quad b = -0.0166, \quad d = -1.0298 \qquad (6.14)$$

At the extremum the functional J is equal to 16.6498.
Experimental and calculated values are as follows:

$x(t_p)$ calc.	1.876	1.99	2.12	2.40	2.65	2.91
$x_e(t_p)$ exp.	1.876	1.99	2.00	2.60	2.35	2.98

A good approximation is obtained and optimal control may be considered. Firstly, let us notice that the model (6.9) which is equivalent to (6.12) has useful applications. For example a patient has the possibility of predicting the variation in his insulin concentration. It is sufficient to measure x_0, the glucose concentration, at time t_0 and to take the previous injections of insulin into account. The formula (6.12) gives the variation in blood glucose level for $t \geq 0$, because the parameters a, b, and d are valid at all times. Therefore, the patient knows empirically whether or not an insulin injection is necessary. In some sense he may be able to achieve an approximate control. For this, only a pocket calculator is necessary.

6.3 Optimal control for optimal therapeutics

The aim of a good medical treatment is to bring the function $x(t)$ to a desired optimal constant A. Using a mathematical formulation, the following problem is set:

Find the c_p and the t_p (time of injection) such that:

$$\int_0^\infty (x(t) - A)^2 dt \quad \text{be minimised.} \qquad (6.15)$$

The function $x(t)$ is given by (6.12) and therefore the

optimal control problem (6.15) is a classical minimization
system where the functional to minimise:

$$\int_0^\infty (x(t) - A)^2 dt$$

explicitly contains the control variables c_p and t_p, where
$p = 0, \ldots, 5$.

A particular case of (6.15) consists of fixing the t_p
(using the physician's experience, for instance) and looking
for the doses c_p. The functional to minimise is given by:

$$J = \int_0^\infty (x(t) - A)^2 dt = \int_0^\infty [x_0 \exp(at)$$

$$+ b/(d-a) \sum_{p=0}^m c_p Y(t-t_p)(\exp(d(t-t_p)) - \exp(a(t-t_p))) - A]^2 dt$$

$$(6.16)$$

A necessary condition for optimality is:

$$\partial J/\partial c_p = 0 \quad , \quad p = 0, \ldots, m \qquad (6.17)$$

which gives a linear algebraic system. Its solution gives
the desired optimal result.

Returning to the general optimization problem where the
c_p and t_p have to be determined, the solution may be found
by using an optimization method (local or global). Another
possibility is to use an iterative algorithm with linear
steps.

(i) First we fix the t_p at reasonable values and J is
minimised with respect to the c_p. For this, the linear
algebraic system:

$$\partial J/\partial c_p = 0 \quad , \quad p = 0, \ldots, m$$

is solved, and the optimal solution:

$$(c_p^*) \quad , \quad p = 0, \ldots, m$$

is obtained.

(ii) Then with the derived values (c_p^*) we look for the
minimum of J according to:

$$\alpha_p = \exp(-dt_p), \quad p = 0, \ldots, m$$

the values of $\exp(-at_p)$ being fixed from the values chosen at the first step. The α_p are determined by solving the linear system:

$$\partial J/\partial \alpha_p = 0, \quad p = 0, \ldots, m \qquad (6.18)$$

Note (α_p^*) the optimal solution.

(iii) Setting:

$$\beta_p = \exp(-at_p), \quad p = 0, \ldots, m$$

the optimum of J with respect to the β_p is calculated by:

$$\partial J/\partial \beta_p = 0, \quad p = 0, \ldots, m$$

which is a linear system giving (β_p^*) the optimal solution.

Then we go back to the first step and the process is iterated until two successive iterations have "almost" the same value.

A more complicated control problem will be considered in the following.

This optimization problem may be solved numerically using the global minimization technique, Alienor, described previously. R. Kamga [45] made the calculation in two cases:

a) when the injection times are fixed, and

b) with times of injection and doses unknown.

The following numerical results were obtained:

Injection times fixed		Times of injection unknown	
experimental doses	optimal doses calculated	optimal times of injection	optimal doses
$c_0 = 1.36$	$c_0 = 0.997$	$t = 0$	$c_0 = 0.1589$
$c_1 = 0.22$	$c_1 = 0.394$	$t = 3$	$c_1 = 2.373$
$c_2 = 1.98$	$c_2 = 0.740$	$t = 13$	$c_2 = 2.332$
$c_3 = 0.53$	$c_3 = 0.962$	$t = 20$	$c_3 = 0.032$
$c_4 = 1.95$	$c_4 = 0.134$	$t = 21$	$c_4 = 0.831$
$c_5 = 0.46$	$c_5 = 1.808$	$t = 22$	$c_5 = 0.482$

In the case where times and doses are unknown, a numerical solution is easily obtained using the Alienor technique.

6.4 Optimal control problem involving several criteria

In many systems at least two criteria are necessary for good optimization. For example, when considering the following compartmental model:

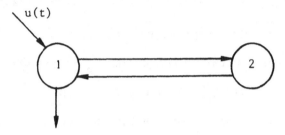

for which the differential equations have already been written, one may introduce the two criteria:

$$\underset{u(t)}{\text{Min}} \ \int_0^\infty (x_1(t) - a)^2 dt \qquad (6.20)$$

$$\underset{u(t)}{\text{Min}} \ \int_0^\infty x_2^2(t) dt \qquad (6.21)$$

The first criterion (6.20) is classical; the second corresponds to the minimization of an undesired effect. The mathematical problem consists of defining some priority to solve (6.20), (6.21). In fact it is not possible to solve them separately because different optimal therapeutics would be obtained, and we want to find a therapeutic which is acceptable for both problems, (6.20) and (6.21).

In connection with the problem of glucose level studied in sections 6.1 and 6.2, another criterion may be added. For example:

$$\sum_{p=0}^m c_p^2 \qquad (6.22)$$

has to be minimised. We may also add the criterion:

$$\underset{c_0,\ldots,c_m}{\text{Min}} \int_0^T (y(t) - B)^2 dt \qquad (6.23)$$

where B is an "optimal" constant associated with the insulin concentration. There are two possibilities for solving a problem such as (6.20),(6.21).

(i) The first [4] consists of building a new unique functional associating the two criteria. For example:

$$\underset{u(t)}{\text{Min}} \ [\int_0^\infty (x_1(t) - a)^2 dt + \lambda \int_0^\infty x_2^2(t) dt] \qquad (6.24)$$

may be considered. In (6.24) λ is a constant term to be chosen as a function of the respective importance of each criterion. For instance, if:

$$\int_0^\infty (x_1(t) - a)^2 dt$$

must be given priority in the minimization, λ must be small. On the other hand, if the second criterion is the most important, λ must be large.   Therefore, there is some uncertainty in the choice of λ ; good values are often indicated by numerical experiments but the technique may be difficult to apply. Alternatively classical optimization techniques are used for solving (6.24) when x_1 and x_2 are explicit functions of u(t). This is the case when considering the compartmental system for which:

$$x_1 = K_1 * u$$
$$\qquad\qquad\qquad\qquad (6.25)$$
$$x_2 = K_2 * u$$

where $*$ is the convolution product.

(ii) The second possibility is to use the hierarchical control approach as follows. Consider a general problem where two functionals J_1 and J_2, continuous, strictly convex and positive, are given. Furthermore they satisfy:

$$\lim_{\|v\| \to \infty} J_i(v) = +\infty$$

Let u* be the point satisfying:

$$\text{Inf } J_1(v) = J_1(u^*) \qquad\qquad (6.26)$$

From previous hypotheses, we know that u* exists and is unique. The optimization of J_1 and J_2 may be achieved by introducing the following problem:

Find u*, v* such that

$$\text{Inf}_{v \in V} \quad J_1(v) = J_1(u^*) \qquad\qquad (6.27)$$

$$\text{Inf}_{\|v-u^*\| \leqslant \rho} \quad J_2(v) = J_2(v^*)$$

where ρ is a fixed number and V is the space where J_1 and J_2 are defined (in practice, it is always R^n). Therefore (6.27) consists of finding a minimum of J_2 in the neighbourhood of a minimum of J_1. The neighbourhood is in fact a sphere of radius ρ and centre u*. To solve (6.27) it is, of course, possible to first find u* using an optimization technique, and then to solve the second optimization problem with constraints. But there is also another possiblity which gives u* and v* in one step.
For this we introduce the functional [8]:

$$J_\epsilon(u,v) = J_1(u) + \epsilon J_2(u + v) + G^2(v)/\epsilon \qquad\qquad (6.28)$$

where $\epsilon > 0$ is a parameter which tends towards zero and where $G(v) = (\|v\| - \rho)^+$ and $f^+(t) = 0$ if $f(t) \leqslant 0$, otherwise $f^+(t) = f(t)$.

The following theorem may be proved.

Theorem 6.1 The functional J_ϵ has a unique minimum on $V \times V$ (denoted by (u_ϵ, v_ϵ)) such that:

$u_\epsilon \to u^*$ and $v_\epsilon \to v^*$ when ϵ tends towards 0. The convergence is in fact a weak convergence which is defined below.

The proof of the existence and uniqueness of (u_ϵ, v_ϵ) is easy and uses the previous results on optimization taking into account the properties of J. (J is continuous and strictly convex). Let us choose v_0 with $\| v_0 \| = \rho$ and $u = v_0$. We have:

$$\text{Inf} \quad J_\epsilon(u,v) = J_\epsilon(u_\epsilon, v_\epsilon) \leqslant J_1(v_0) + \epsilon J_2(2v_0) + G^2(v_0)/\epsilon$$

giving:

$$J_1(u_\epsilon) \leqslant J_1(u_\epsilon) + \epsilon J_2(u_\epsilon + v_\epsilon) + G^2(v_\epsilon)/\epsilon$$

$$\leqslant J_\epsilon(u_\epsilon, v_\epsilon) \leqslant J_1(v_0) + \epsilon J_2(2v_0) \qquad (6.29)$$

Now $\lim\limits_{\|v\| \to \infty} J_1 = \infty$ gives, because of the inequality (6.29):

$$\| u_\epsilon \| \text{ is bounded (independently of } \epsilon) \qquad (6.30)$$

But the following inequality holds:

$$J_1(u_\epsilon) + \epsilon J_2(u_\epsilon + v_\epsilon) \leqslant J_1(u^*) + \epsilon J_2(u^* + v_0) \qquad (6.31)$$

As $\lim\limits_{\|v\| \to \infty} J_2(v) = \infty$ we deduce $\| u_\epsilon + v_\epsilon \|$ bounded because

(6.31) gives:

$$\epsilon J_2(u_\epsilon + v_\epsilon) \leqslant J_1(u^*) - J_1(u_\epsilon) + \epsilon J_2(u^* + v_0) \leqslant \epsilon J_2(u^* + v_0)$$
$$(6.32)$$

The conclusion follows:

$$\| v_\epsilon \| \text{ is bounded, and therefore:}$$

$$u_\epsilon \to u_1 \text{ weakly in V}, \qquad v_\epsilon \to v_1 \text{ weakly in V} \qquad (6.33)$$

where V is assumed to be a Hilbert space.

But, from (6.31):

$$J_1(u_\epsilon) \leqslant J_1(u^*) - \epsilon J_2(u_\epsilon + v_\epsilon) + \epsilon J_2(u^* + v_0)$$

Taking the lim in this last relation gives (when $\epsilon \to 0$):

$$J_1(u_1) \leqslant J_1(u^*)$$

The uniqueness of u^*, the minimum of J_1 implies $u_1 = u^*$.

On the other hand, we showed that:

$$J_2(u_\epsilon + v_\epsilon) \leqslant J_2(u^* + v_0) \quad \text{for all } v_0 \text{ with } \|v_0\| = \rho$$

Considering the $\underline{\lim}$ of this expression when $\epsilon \to 0$ gives:

$$J_2(u^* + v_1) \leqslant J_2(u^* + v_0) \quad \text{for all } v_0 \text{ with } \quad \|v_0\| = \rho \quad (6.34)$$

This implies $v_1 = v^*$. Indeed v^* belongs to the boundary of $[v \mid \|v\| \leqslant \rho]$. If it were not true, v^* would be the global optimum of $J_2(v)$ and the problem (6.23) would become:

$$\text{Inf } J_1(v) = J_1(u^*) \quad , v \in V$$
$$\quad (6.35)$$
$$\text{Inf } J_2(v) = J_2(v^*) \quad , v \in V$$

and this can be excluded because we do not have a hierarchical control system.

The relation (6.34) shows that $v_1 = v^*$ and consequently the sequences u_ϵ and v_ϵ weakly converge to u^* and v^* respectively. In fact we only showed that two subsequences of u_ϵ and v_ϵ converge, but it is easy to demonstrate the convergence of the full sequences. To suppose the contrary leads to a contradiction.

<u>Remark</u>   More generally one may consider optimization problems with constraints, as in the following:

$$\begin{array}{c} \text{Inf} \quad J_1(v) = J_1(u^*) \\ v \in K_1 \end{array}$$
$$\quad (6.36)$$
$$\begin{array}{c} \text{Inf} \quad J_2(v) = J_2(v^*) \\ v \in K_2 \\ \|v - u^*\| \leqslant \rho \end{array}$$

The associated unique functional could be:

$$J_\epsilon^{1,2}(u,v) = J_1(u) + J_2(u + v) + G^2(v)/\epsilon + H_1^2/\epsilon + H_2^2/\epsilon \quad (6.37)$$

where $H_1(v) = G_1(v)^+$, $H_2(v) = G_2(v)^+$

and the convex K_1, K_2 defined by:

$$K_1 = [v \mid G_1(v) \leqslant 0] \quad , \quad K_2 = [v \mid G_2(v) \leqslant 0] \qquad (6.38)$$

Furthermore the J_ϵ considered could be replaced by:

$$J_{\epsilon_1, \epsilon_2} = J_1(u) + \epsilon_1 J_2(u + v) + G^2(v)/\epsilon_2 \qquad (6.39)$$

where $\epsilon_1 > 0$, $\epsilon_2 > 0$ are two parameters tending towards zero.

The previous demonstration of convergence may be adapted without difficulty.

CHAPTER 7

INTEGRAL EQUATIONS IN BIOMEDICINE

7.1 Compartmental analysis

In the previous chapters which were devoted to compartmental analysis we saw that the input (control) $u(t)$ and the quantity, or concentration, in each compartment was given by an integral equation (linear) of the convolution type. This linear integral equation may be written as follows:

$$x_i(t) = K_i(t) * u(t) \quad , \ x_i = \text{concentration in } i; \qquad (7.1)$$

K_i is the kernel [69] and is calculable from the exchange parameters k_{ij}. The function $u(t)$ is the input variable and, in our studies, is also the control variable.

Moreover, when considering a non-linear complex system where only the time is taken into account, the formula (7.1) may be generalized. For example when considering a relation between blood concentration and drug effect, the following was chosen for a beta-blocker acting on the heart:

$$y(t) = K(t) * g(x(t)) \qquad (7.2)$$

where $K(t)$ is a kernel function to be identified from experimental data and $g(t)$ is a non-linear function which also has to be identified. For the beta-blocker (Servier Laboratories) being studied, the function g was taken equal to a two- or three-degree polynomial function. The functions $K(t)$ and $g(t)$ having their supports on $(0,\infty)$, (7.2) is an integral equation of the Volterra type [69] and is of the first kind. More sophisticated equations may be introduced for studying more complex biological phenomena. The next equation:

$$y(t) = \int_0^T K(t,s) \ g(u(s)) \ ds + \lambda u(t) \qquad (7.3)$$

152

may be considered. The kernel, K, the constant λ, and the
function g have to be identified from the available data.
In (7.3), for example, y(t) is a variable appearing in the
optimised criterion; u(t) is the control variable. This
formulation is very general and, in practice, can be adapted
to every non-linear biological system provided that the only
variable taken into account is time. To complete this
study let us introduce an associated control problem.

Find u(t) such that:

$$\int_0^T (y(t) - A)^2 dt \text{ be minimised} \qquad (7.4)$$

where y(t) is a solution of (7.3) with $\lambda = 0$

Choosing g(u(t)) as a new control variable, we have:

$$\int_0^T (y(t) - A)^2 dt = \int_0^T [\int_0^T K(t,s)g(u)ds - A]^2 dt \qquad (7.5)$$

Setting:

$$g(u(t)) = \sum_{i=1}^n a_i \theta_i(t) \qquad (7.6)$$

where $\theta_i(t)$ are chosen functions forming a basis in the
space of the functions u.
    Putting (7.6) into (7.5) involves a new optimization
problem with respect to the a_i, i = 1, .. n   that is:

$$\underset{a_1,\ldots,a_n}{\text{Min}} \int_0^T [\int_0^T K(t,s) \sum_{i=1}^n a_i \theta_i(u).du - A]^2 dt = \underset{a_1,\ldots,a_n}{\text{Min}} \quad J \qquad (7.7)$$

whose solution is given by the necessary conditions for an
optimum:

$$\partial J/\partial a_1 = \partial J/\partial a_2 = \ldots = \partial J/\partial a_n = 0 \qquad (7.8)$$

In fact, the system (7.8) is an algebraic linear system and,
therefore, the optimal solution $a_1^*, \ldots, a_n^*$ satisfying:

$$\underset{a_1,\ldots,a_n}{\text{Min}} \qquad J = J \ (a_1^*, \ . \ . \ ,a_n^*) \qquad\qquad (7.9)$$

is easy to obtain.

Suppose that $g(u)$ has an inverse function g^{-1}, then the optimal control u^* can be deduced. Indeed we saw that:

$$g(u^*(t)) = \sum_{i=1}^{n} a_i^* \theta_i(t) \qquad\qquad (7.10)$$

and therefore $u^*(t)$ is given by the formula:

$$u^*(t) = g^{-1}[\ \sum_{i=1}^{n} a_i^* \theta_i(t)] \qquad\qquad (7.11)$$

Of course, a numerical approximation may be used for g^{-1} in the case where an explicit relation for g^{-1} is not available. This technique is also valid for a convolution relation as in (7.2). The problem is more difficult when $\lambda \neq 0$ but a numerical method is possible using a linearization technique for the solution of integral equations of Fredholm type [69].

This type of modelling does not lead to a knowledge model. Such a model will be developed in the following.

7.2 Integral equations from biomechanics

We shall present a rather general model arising when substances are passing through tubes. Let us consider a rigid tube of length L and radius R. In this tube blood, containing a dye, moves with constant speed v. At $x = 0$, the dye is injected and mixed. The mathematical problem consists in finding a relation between the input concentration (at $x = 0$) and the output concentration (at $x = L$). The tube (called a catheter in surgery) is supposed impermeable. It may be used, for instance, to measure ventricular volumes. Let us note $C(x,t)$ the dye concentration at time t and abscissa x. For exchanges during respiration one may use a mass balance (see Chapter 1).

The next equation follows directly:

$$\partial C/\partial t = - v. \ \partial C/\partial x \qquad (7.12)$$

Its solution is easy and leads to the relation:

$$C(L,t) = C(0,t - L/v) \qquad (7.13)$$

where $C(0,t)$ is the input concentration and $C_L(t) = C(L,t)$ the output concentration of the dye. (7.13) implies that the output is displaced in time by L/v relative to the input. This result is false from the experimental point of view. The conclusion is that v is not an absolute constant; it depends on r, the variable in polar co-ordinates. In most cases one uses the formula:

$$v(r) = (2Q/\pi R^2) \ (1 - r^2/R^2) \qquad (7.14)$$

where Q is the blood flow. This relation (7.14) corresponds to a laminar flow. For the blood, it is only a good approximation. Now we are going to build a knowledge model. Note that $C_0(t)$ is the input concentration and $C_L(t)$ is the mean of the output concentration. Let us consider a section element (at $x = L$) included between r and $r + dr$. The area of the section is approximately $2\pi r dr$. In this small element $v(r)$ may be considered as a constant independent of x, r, and t. The matter flowing through this element is equal to:

$$2\pi r v(r) dr \qquad (7.15)$$

It is the product of speed and area.

But from (7.12) and (7.13) the concentration in the ring-shaped element (at $x = L$) is equal to:

$$C_0(t - L/v(r)) \qquad (7.16)$$

and therefore the flow of dye through our small element at $x = L$, is given by:

$$2\pi r v(r) \ C_0(t - L/v(r)) dr \qquad (7.17)$$

By integrating (7.17) it becomes possible to calculate the total dye flow at $x = L$. We obtain:

$$\int_0^R 2\pi r v(r) \ C_0(t - L/v(r)) dr \qquad (7.18)$$

This is the first formula for the total flow. Another method allows us to determine this flow. Indeed at $x = L$, the dye flow is equal to:

$$QC_L(t) \quad \text{where} \quad C_1(t) = (1/R) \int_0^R C(L,r)dr \qquad (7.19)$$

Comparing (7.19) and (7.18) gives the formula we are looking for:

$$C_L(t) = (2\pi/Q) \int_0^R rv(r)C_0(t - L/v(r))dr \qquad (7.20)$$

which is a linear integral equation of Fredholm type (first kind) [1].

Some changes of variables transform (7.20) into Volterra equations of the second type:

$$D_0(t) = (L/v_{ax})C_L'(t)$$

$$+ (3L/2v_{ax}) \int_{1/v_{ax}}^t [D_0(w)/((t - w + L/v(0))^4)]dw \qquad (7.21)$$

where $v_{ax} = 2Q/\pi R^2 = v(0)$, $D_0(w) = C_0(w - L/v(0))$

Then classical results show [69] that (7.21) has a unique solution when:

$$C_L(t) \in C^{(1)} (0,\infty)$$

The formulae (7.20) or (7.21) may be used in practice to determine ventricular volumes using the expression:

$$V = [q \int_0^\infty tC_0(t)dt]/(\int_0^\infty C_0(t)dt)^2 \qquad (7.22)$$

where q is the quantity of dye injected.

Numerical methods for the solution of (7.20) and (7.21) will be developed later, but already we can notice the following.

Remark Non-linear integral equations arise when considering complex flows. For example the following relation between C_0 and C_L:

$$c_L(t) = c_0(t) + c_0^3(t - L/v(r))   \qquad (7.23)$$

leads to the integral equation (7.24):

$$c_L(t) = Kc_0(t) + 2\pi/Q \int_0^R rv(r)c_0^3(t - L/v(r))dr \quad (7.24)$$

where K is a constant that may be calculated from (7.23) using the previous reasoning.

7.3 Other applications of integral equations

a) The approach of Volterra to the problem of the growth of populations needs integral equations. This is especially so when the influence of heredity is taken into account when studying population growth. A classical differential system, as in compartmental analysis, is not sufficient and an integro-differential equation is found [52].

$$(1/y)dy/dt = a + by + \int_c^t K(t,s)y(s)ds \qquad (7.25)$$

where $y(t)$ is the population size at time t, and a and b are constants depending on the population. In the case of two competing populations, one preying on the other, Volterra introduced the following system:

$$(1/x)\dot{x} = \quad a - by - \int_{-\infty}^t K_1(t - s)y(s)ds$$

$$\qquad (7.26)$$

$$(1/y)\dot{y} = -\alpha y + \beta x + \int_{-\infty}^t K_2(t - s)x(s)ds$$

where $x(t)$ is the size of the first population and $y(t)$ corresponds to the second one.

b) Differential systems may be transformed into integral equations and therefore solved using numerical techniques which are available for integral equations.

For example, let us consider the differential equation:

$$d^2w/ds^2 + a^2sw = 0 \qquad 0 \leqslant s \leqslant 1 \qquad (7.27)$$

with boundary conditions expressed as:

$$s = 0 \quad , \qquad dw/ds = 0$$
$$\qquad\qquad\qquad\qquad\qquad\qquad\qquad (7.28)$$
$$s = 1 \quad , \qquad\qquad w = 0$$

This equation plays a role in buckling loads [52]. Integrating (7.27) twice with respect to s, integrating by parts and making use of the boundary conditions (7.28) we obtain:

$$w(s) = a^2 \int_0^1 (1 - t)tw(t)dt - a^2 \int_0^s (s - t)tw(t)dt \qquad (7.29)$$

which may be rewritten as:

$$w(s) = \lambda \int_0^1 k(s,t)w(t)dt \qquad (7.30)$$

where the kernel k(s,t) is given by:

$$k(s,t) = \begin{cases} (1 - t)t & \text{if } s \leqslant t \leqslant 1 \\ (1 - s)t & \text{if } 0 \leqslant t \leqslant s \end{cases} \qquad (7.31)$$

We saw previously that compartmental models may be transformed in integral convolution equations and therefore modelling by integral equations is a very general representation of biological phenomena. Even partial differential equations may be reduced to integral equations with some difficulty [1].

c)   Integral equations are often used in biomechanics. For example the fundamental fluid transport mechanism in a joint cavity [41] and the deformation of bearing surfaces are important for the understanding of the biomechanical processes involved in the physiology of normal and pathological synovial joints. The study of elasto-rheo-dynamic lubrication of a self-acting dynamically loaded bearing which employs a porous, elastic material (articular cartilage) and a visco-elastic non-linear lubricant (synovial fluid) leads to a non-linear integral equation [41]:

$$u(x) + \lambda \int_a^b g(u(t)).K(x,t)dt = f(x) \qquad (7.32)$$

where g, K, f and λ are obtained from physical reasoning. The function u(x) has to be determined and corresponds to the thickness of the film of synovial fluid [41]. In practice, g(u) has the form $g(u) = 1/u^3$ and therefore is strongly non-linear. In the following we shall develop numerical methods to solve (7.32).

d) Oxygen diffusion can also involve non-linear integral equations. This fact arises when analysing oxygen diffusion in a spherical cell which has non-linear Michaelis-Menten oxygen uptake kinetics. A recent paper is devoted to this approach [68]. Following Lin [68], the unsteady-state oxygen diffusion in a spherical cell may be represented in the following non-dimensional form:

$$\partial C/\partial \tau = \partial^2 C/\partial R^2 + (2/R)\partial C/\partial R - \alpha C/(C + K_m) \qquad (7.33)$$

with the boundary and initial conditions:

$$R = 0 \quad , \quad \partial C/\partial R = 0$$

$$R = 1 \quad , \quad \partial C/\partial R = H(1-C)$$

$$\tau = 0 \quad , \quad C = 0$$

where C measures the oxygen tension, R the radial co-ordinate, τ the time variable, α the reaction rate, H the permeability of the membrane and K_m the Michaelis-Menten constant. In [68] the authors show how an integral equation method can be used to find good approximate solutions for non-linear diffusion equations whether the problem is steady or unsteady.

The non-linear differential equation (7.33) and the initial and boundary conditions are transformed into an integral equation by multiplying (7.33) by R^2 and integrating the result from 1 to any R. We obtain:

$\partial C/\partial R = (1/R^2)H[1 - C(1)]$

$$+ (1/R^2)[\int_1^R \rho^2(\partial C/\partial \tau)d\rho + \int_1^R (\alpha c\rho^2/(C + K_m))d\rho]$$

Integrating again over the same interval gives:

$$C(R) = 1 + (1 - 1/H)\int_0^1 \rho^2(\partial C/\partial \tau + \alpha c/(C + K_m))d\rho$$

$$- 1/R\int_0^R \rho^2(\partial C/\partial \tau + \alpha c/(C + K_m))d\rho \qquad (7.34)$$

$$+ \int_1^R \rho(\partial C/\partial \tau + \alpha c/(C + K_m))d\rho$$

For further details see [68].

In the case of the steady state problem we have $\partial C/\partial \tau = 0$ and therefore:

$$C(R) = 1 + (1 - 1/H)\int_0^1 \rho^2(\alpha c/(C + K_m))d\rho$$

$$- 1/R\int_0^R \rho^2(\alpha c/(C + K_m))d\rho \qquad (7.34')$$

$$+ \int_1^R \rho(\alpha c/(C + K_m))d\rho$$

(7.34') is of course a classical integral equation and may be solved using numerical methods described in Chapter 8.

A. Caron, in a non-published paper, tried to solve (7.34') and (7.34) with a numerical technique involving a regularity constraint. (See Chapter 8). Good results were obtained in a very short time, thus proving the validity of the method. Equation (7.34) is integro-differential. The

method of Caron can be used but, before using it, it is
necessary to discretise the partial derivative with respect
to τ. For example, an approximation of $\partial C/\partial \tau$ may be:

$$\partial C/\partial \tau \ = \ (C(R, \ \tau+\Delta\tau \ ) \ - \ C(R,\tau))/\Delta\tau \quad \text{with } \Delta\tau \quad \text{small.}$$

Numerical results and graphs may be found in [68].

CHAPTER 8

NUMERICAL SOLUTION OF INTEGRAL EQUATIONS

8.1 Linear integral equations

This is the most simple problem because we obtain a linear algebraic system. We shall not look at the problems of existence and uniqueness of integral equations in this book. There are many good works on these subjects [1], [69]. Let us consider the following general linear integral equation:

$$u(x) + \int_0^1 K(x,t)u(t)dt = f(x) \qquad (8.1)$$

where f and K are given. The problem is to determine $u(t)$ numerically such that (8.1) is satisfied. For that we shall use the approximation:

$$u(x) = \sum_{p=1}^{n} c_p r_p(x) \qquad (8.2)$$

where the functions r_p are known and $\left\{ r_p(x) \right\}_{p=1}^{\infty}$ is a basis in the set of u's.

Putting (8.2) into (8.1) gives:

$$\sum_{p=1}^{n} c_p r_p(x) + \sum_{p=1}^{n} c_p \int_0^1 K(x,t)r_p(t)dt = f(x) \qquad (8.3)$$

In (8.3) the expressions $K_p(x) = \int_0^1 K(x,t)r_p(t)dt$ are calculable (either explicitly or numerically).

If we want to find a solution of (8.1) for $x \in [0,1]$ it is possible to write (8.3) with $x = qh$, $q = 1, \ldots ,n$ and $nh = 1$.

The following algebraic system is thus obtained:

$$\sum_{p=1}^{n} c_p r_p(qh) + \sum_{p=1}^{n} c_p K_p(qh) = f(qh) \qquad q = 1, \ldots ,n \qquad (8.4)$$

which is linear and has as many equations as unknowns. Using a classical numerical method (Gauss, ..) the c_p may be calculated and we obtain the approximate solution:

$$u(x) = \sum_{p=1}^{n} c_p r_p(x) \qquad (8.5)$$

The $r_p(x)$ was chosen previously and can be a polynomial, exponential or spline function.

For solving (8.1) an optimization technique can also be tried. Introduce the following functional J, such that:

$$J(c_p) = \int_0^1 (\sum c_p r_p(x) + \sum c_p K_p(x) - f(x))^2 dx \qquad (8.6)$$

the solution c_p is obtained by minimising J. But the functional in (8.6) is quadratic and thus the necessary conditions for a minimum:

$$\partial J/\partial c_p = 0 \quad , p = 1, \ . \ . \ ,n \qquad (8.7)$$

give a linear algebraic system of n equations and n unknowns. The main interest of this last method is that it takes into account all the values of x belonging to [0,1].

Remark For equations of the Volterra type such as the following:

$$u(t) + \int_0^t K(x,t)u(x)dx = f(t) \qquad (8.8)$$

simpler numerical methods may be proposed. Choosing a step h, which is small, for discretising the time interval $(0,\infty)$, and with t = h, one obtains:

$$u(0) = f(0)$$

$$\qquad (8.9)$$

$$u(h) + \int_0^h K(x,h)u(x)dx = f(h)$$

But: $\int_0^h K(x,h)u(x)dx \simeq hK(0,h)u(0)$ because h is small.

(8.9) gives:

$$u(h) = f(h) + hK(0,h)u(0) \qquad (8.10)$$

and $u(h)$ is therefore known. the process can be iterated as follows:

Taking $t = 2h$ in (8.8) gives the functional relation:

$$u(2h) + \int_0^h K(x,2h)u(x)dx + \int_h^{2h} K(x,2h)u(x)dx = f(2h)$$

Approximating the integrals as before leads to:

$$u(2h) + hK(0,2h)u(0) + hK(h,2h)u(h) = f(2h) \qquad (8.11)$$

which gives $u(2h)$ as a function of known quantities and of $u(0)$ and $u(h)$. Then by use of an explicit formula we shall calculate $u(3h), \ldots, u(nh), \ldots$ In this case we have an explicit formula and do not need to solve the linear algebraic system.

8.2 Numerical techniques for non-linear integral equations

Let us consider the general equation:

$$u(x) = \int_a^b K(x,t)g(u(t))dt + f(x) \qquad (8.12)$$

where K, g and f are given such that (8.12) has a unique solution in well chosen spaces [1], [69]. The methods used for the linear case may, of course, be adapted. For example, by setting:

$$u(x) = \sum_{p=1}^{n} c_p r_p(x) \qquad (8.13)$$

and putting this expression in (8.12) gives the non-linear algebraic system:

$$\sum_{p=1}^{n} c_p r_p(qh) = \int_a^b K(qh,t)g(\sum_{p=1}^{n} c_p r_p(t))dt + f(qh) \qquad (8.14)$$

where $q = 1, \ldots ,n$

This system is obtained by setting $x = qh$.

To solve (8.14) a positive functional has to be introduced:

$$J = \sum_{q=1}^{n} [\sum c_p r_p(qh) - \int_a^b K(qh,t)g(\sum c_p r_p(t))dt - f(qh)]^2 \qquad (8.15)$$

Unfortunately J is not a polynomial and therefore the previous simple methods cannot be used. An optimization technnique (local or global) has to be proposed for determining the unknown coefficients c_p. More simple methods are possible using some linearization techniques.

8.2.1   NUMERICAL   SOLUTION   USING   A   SEQUENCE   OF   LINEAR
INTEGRAL EQUATIONS

Starting with the integral equation (8.12) we shall build a sequence (u_n, v_n) as follows:
$u_0(x)$ is the solution of:

$$u_0(x) = \int_a^b K_0(x,t)g(u_0(t))dt + f(x) \qquad (8.16)$$

where $K_0(x,t)$ is an approximation of K which can be made (for instance K_0 = constant). In fact K_0 has to be chosen so that the calculation of u_0 in (8.16) is easy.

Setting: $R(x,t) = K - K_0$

$$r(x) \quad = \int_a^b R(x,t)g(u_0(t))dt \qquad (8.17)$$

it is easy to show that $v(x) = u(x) - u_0(x)$ is a solution of

$$v(x) = r(x) + \int_a^b K(x,t)(g(u) - g(u_0))dt \qquad (8.18)$$

Note that $r(x)$ is known when u_0 is calculated. Let us set:

$$g(u) - g(u_0) = (u - u_0)G(u_0,u) = G(u_0,u)v(t) \qquad (8.19)$$

The following integral equation results from (8.17),(8.18) and (8.19):

$$v(x) = r(x) + \int_a^b K(x,t)G(u_0,u)v(t)dt \qquad (8.20)$$

Now the numerical method is based on an approximation of $G(u_0,u)$. To be more precise, the following algorithm is defined. Let:

$$u_{n+1} = u_0 + v_n, \qquad u_1 = u_0 + v_0$$

u_1 is found by solving the integral equation:

$$v_0(x) = r(x) + \int_a^b K(x,t)G(u_0,u)v_0(t)dt \qquad (8.21)$$

In fact, (8.21) is a <u>linear</u> integral equation and so the previous methods may be used.

More generally $v_n(x)$ is obtained by solving the linear integral equation:

$$v_n(x) = r(x) + \int_a^b K(x,t)G(u_0,u_n)v_n(t)dt \qquad (8.22)$$

where u_n has been calculated at the step $(n - 1)$.

In [16] the convergence of this method is proved under the following hypothesis:

$$|G(u_0,u)| \leqslant G < +\infty \ , \ |K(x,t)| \leqslant k < \infty \text{ and } r(x) \in H^1(a,b)$$

$$\text{with } H^1(a,b) = \left\{ [v | v \in L^2(a,b), \ v' \in L^2(a,b)] \right\}$$

Numerical experiments show that, in practice, the solution is obtained in three or four iterations. The method can also be used when uniqueness is not assured, but in this case a particular solution is found [17].

8.2.2 A DISCRETISED TECHNIQUE

This method uses ideas already developed in [16]. We shall discuss the technique for the following non-linear equation:

$$u(x) + \int_0^1 K(x,t)u^3(t)dt = f(x) \qquad (8.23)$$

This can be replaced by the equivalent relation:

$$u(x) + \sum_{j=0}^{N-1} \int_{jh}^{(j+1)h} K(x,t)u^3(t)dt = f(x) \qquad (8.24)$$

where N is an integer such that Nh = 1 with h tending towards 0. Taking x = ih in (8.24) leads to:

$$u(ih) + \sum_{j=0}^{N-1} \int_{jh}^{(j+1)h} K(ih,t)u^3(t)dt = f(ih) \qquad (8.25)$$

with i = 0, . . ,N

On (jh,(j+1)h) with h small, the function $g(u) = u^3$ which is non-linear may be approximated by:

$$u^3(jh) + 3u^2(jh)(u(j+1)h) - u(jh)) \qquad (8.26)$$

In fact we used a linearization in a neighbourhood of t = jh. If the general case with g(u) instead of u^3 in equation (8.23) is considered, g(u) would be replaced by $g(u_0) + (u - u_0)g'(u_0)$.
The following algorithm can be proposed:

$$u_n(ih) + \sum_{j=0}^{N-1} hK(ih,(j+1)h)[u_{n-1}^3(jh)$$

$$+ 3u_{n-1}^2(jh)(u_n((j+1)h) - u_{n-1}(jh))] = f(ih) \qquad (8.27)$$

i = 0, . . ,N ;   j = 0, . . ,N-1 ;   n = 1,2, . . .

The $u_0(jh)$, j = 0, . . ,N-1, are fixed (to f(jh) for example) or calculated.
For n fixed the algebraic system (8.27) is <u>linear</u>. The unknowns are the $u_n(jh)$. The numerical solution of (8.27) gives an approximate solution of (8.23). When n → ∞ the convergence of the iterative process can be proved.
Notice that the discretised system in (8.27) is not

unique. The discretization depends on approximations associated with the integrals:

$$\int_{jh}^{(j+1)h}$$

and with $u^3(t)$ or $g(u)$. It is even possible to reduce (8.23) to the solution of a sequence of linear equations. With this objective let us introduce the iterative form:

$$u_n(ih) + \sum_{j=0}^{i-1} \int_{jh}^{(j+1)h} K(ih,t)u_n^3(t)dt$$

$$+ \sum_{j=i}^{N-1} \int_{jh}^{(j+1)h} K(ih,t)u_{n-1}^3(t)dt = f(ih) \qquad (8.28)$$

$$\text{for } i = 1, \ldots ,N-1$$

The following approximation, valid if h is small, is used in (8.28):

$$\int_{jh}^{(j+1)h} K(ih,t)u_n^3(t)dt = hK(ih,(j+1)h)[u_{n-1}^3(jh)$$

$$+ 3u_{n-1}^2(jh)(u_n((j+1)h) - u_{n-1}(jh))] \qquad (8.29)$$

Therefore (8.28) becomes:

$$u_n(ih) + \sum_{j=0}^{i-1} hK(ih,(j+1)h)[u_{n-1}^3(jh)$$

$$+ 3u_{n-1}^2(jh)(u_n((j+1)h) - u_{n-1}(jh))]$$

$$+ \sum_{j=i}^{N-1} \int_{jh}^{(j+1)h} K(ih,t)u_{n-1}^3(t)dt = f(ih), \quad i = 1, \ldots ,N$$

$$(8.30)$$

The iterative formula (8.30) gives $u_n(h)$, $u_n(2h)$, . . . by solving a sequence of linear equations. As said previously, the approximation associated with:

$$K(ih,t)u_n^3(t)dt$$

may be improved. Of course, in (8.30) $u_n(0)$ has to be
fixed. The reference [17] gives possible choices. This
technique was used on concrete examples and gave good
numerical results [17].

8.2.3 AN ITERATIVE DIAGRAM WITH REGULARITY CONSTRAINTS [9]

This ingenious technique discovered by A. Caron [10] is very
powerful and allows the solution of complex concrete
problems from biomechanics (synovial joints, . .) and other
areas such as electromagnetism, . . .

Returning to the equation:

$$u(x) = \int_a^b K(x,t)g(u(t))dt + f(x) \qquad (8.31)$$

a quadratic formula (after Gauss) may be used to approach
$\int_a^b K(x,t)g(u(t))dt$ by $\sum_{i=1}^n C_i K(x,b_i)g(u(b_i))$ where the b_i
belongs to $[a,b]$. Let us consider the sequence:

$$(a_0, b_1, a_1, \cdot \cdot \cdot, b_n, a_n) \qquad (8.32)$$

where the a_i are arbitrarily chosen with the constraint
$b_i < a_i < b_{i+1}$. The non-linear algebraic system derived
from (8.31) is then introduced:

$$u(a_j) - \sum_{i=1}^n C_i K(a_j,b_i)g(u(b_i)) = f(a_j)$$

$$(8.33)$$

$$u(b_k) - \sum_{i=1}^n C_i K(b_k,b_i)g(u(b_i)) = f(b_k)$$

(8.33) may be linearised using the approximation:

$$u(b_k) = \sum_{l=0}^n u(a_l)T_l(b_k) \qquad (8.34)$$

where $T_1(x)$ is a well-defined polynomial (Tchebycheff, Legendre, . .) which is used to approximate a function given at some points. In our case the function is defined at the points $(a_1, u(a_1))$. The relation (8.34) is the regularity constraint indicated in the title.

Putting (8.34) into (8.33) gives rise to a linear system having $(2n + 1)$ unknowns (the $u(a_j)$ and $g(u(b_i))$) and $(2n + 1)$ relations. A unique solution is obtained if it has a non-zero determinant. It is an approximate solution of the integral equation (8.31). A convergence theorem can be proved [10]. The proof is rather more complicated than the numerical algorithm.

8.3 Identification and optimal control using integral equations

When a biological phenomenon is described by the non-linear integral equation:

$$u(x) = \int_a^b K(x,t)g(u(t))dt + f(x) \qquad (8.35)$$

the identification of the model means that $K(x,t)$ has to be calculated from the knowledge of $u(x)$, $f(x)$ and $g(t)$. In some cases even g may be partially unknown, while in others K is known but g has to be identified.

For example, if only K has to be identified we set:

$$K(x,t) = \sum_{p=1}^n a_p r_p(x) s_p(t) \qquad (8.36)$$

where r_p and s_p are known functions chosen to give a good approximation to the two-variable function. Then (8.35) can be solved by introducing the functional:

$$J = \int_a^b [u(x) - \int_a^b \sum a_p r_p(x) s_p(t) g(u(t))dt - f(x)]^2 dt \quad (8.37)$$

Then a minimization technique is necessary to find:

$$\underset{a_1, \cdots, a_n}{\text{Min}} \quad J \qquad (8.38)$$

J given by (8.37) is a quadratic function and therefore the necessary condition for the optimum:

$$\partial J/\partial a_p = 0 \quad , \quad p = 1, \ldots ,n \tag{8.39}$$

gives an algebraic linear system whose solution provides (by (8.36)) an approximated solution of (8.35).

8.4 Optimal control and non-linear integral equations

In the chapter devoted to optimal control associated with linear compartmental systems we developed techniques involving linear convolution equations. In section (7.1) we treated a particular non-linear control problem involving an integral equation of the first kind. Let us consider a general non-linear integral equation as follows:

$$u(x) + \int_a^b K(x,t)g(u(t))dt = y(x) \tag{8.40}$$

and the associated optimization system:

$$\underset{u(x)}{\text{Min}} \int_a^b (y(x) - A)^2 dx \tag{8.41}$$

In (8.41) the function $y(x)$ can represent an effect or a concentration of a drug; $u(x)$ may be the injected amounts. The functions K and g are known.
 Substituting equations (8.40) into (8.41) gives:

$$\underset{u(x)}{\text{Min}} \int_a^b [u(x) + \int_a^b K(x,t)g(u(t))dt - A]^2 dx = \underset{u(x)}{\text{Min}} J \tag{8.42}$$

with:
$$J = \int_a^b [u(x) + \int_a^b K(x,t)g(u(t))dt - A]^2 dx \tag{8.43}$$

J is a non-linear function and therefore an optimization technique can be used after discretization. Choosing the same approximation as in Section 8.2.3 we obtain the discretised functional J_n associated with J by:

$$J_n = \sum_{j=1}^{n} [u(a_j) + \sum_{i=1}^{n} C_i K(a_j, b_i) g(u(b_i)) - A]^2$$

(8.44)

$$+ \sum_{k=1}^{n} [u(b_k) - \sum_{i=1}^{n} C_i K(b_k, b_i) g(u(b_i)) - A]^2$$

J_n must be minimised in terms of $u(a_j)$, $u(b_k)$ and $g(u(b_i))$. In the form (8.44) J_n is strongly non-linear but a regularity constraint can be introduced as before:

$$u(b_k) = \sum_{l=0}^{n} u(a_l) T_l(b_k)$$

(8.45)

with T_l = known polynomial.

Putting (8.45) into (8.44) gives a quadratic functional depending on the unknowns $u(a_j)$, $g(u(b_i))$.

The optimal solution is obtained by writing:

$$\partial J_n / \partial u_i = \partial J_n / \partial g_i = 0, \quad i = 1, \ldots, n$$

(8.46)

with $u_i = u(a_i)$, $g_i = g(u(b_i))$.

The advantage is the linearity of (8.46), giving 2n relations and 2n unknowns. The numerical solution gives an approximate solution of (8.41).

Remark Without the regularity constraint (8.45) it would not be possible to find a good approximation by minimising J_n. In fact, $u(b_k)$ would also be unknown and the system (8.33) would have an infinity of solutions ((3n + 2) unknowns and only (2n + 1) equations). Furthermore some weights may be added in formula (8.44).

CHAPTER 9

PROBLEMS RELATED TO PARTIAL DIFFERENTIAL EQUATIONS

9.1 General remarks

In Chapter 1 on modelling, we saw a general model leading to a partial differential system of equations. These equations describe gas exchanges in a human organism. Similar models are obtained when considering phenomena based on space and time. For instance, another example leading to partial differential equations comes from enzyme biochemistry, where an enzyme E added to a substrate S [48] gives a product in accordance with:

$$E + S \; \underset{k_{-1}}{\overset{k_1}{\rightleftharpoons}} \; ES \qquad (9.1)$$

Then the complex ES is decomposed, giving a product P and the enzyme:

$$ES \; \overset{k_2}{\longrightarrow} \; E + P \qquad (9.2)$$

The last reaction is irreversible.

To find the mathematical equations corresponding to our variables, a mass balance is done using the following equations where [S], [E], [ES] and [P] represent the respective concentrations of S, E, ES and P:

$$[\dot{E}] = - k_1[E][S] + (k_{-1} + k_2)[ES]$$

$$[\dot{S}] = - k_1[E][S] + k_{-1}[ES]$$

$$[\dot{ES}] = k_1[E][S] - (k_{-1} + k_2)[ES] \qquad (9.3)$$

$$[\dot{P}] = k_2[ES]$$

to which the relation:

$$[E] + [S] = [E]_0 \qquad (9.4)$$

173

may be associated.

In (9.4) $[E]_0$ represents the total concentration of active sites [67]. Chemists generally use one of the two following hypotheses:

a) <u>Michaelis-Menten hypothesis</u>

This supposes that the speed of transformation of ES into P and E is weak in comparison to the speed of the reconversion of ES into E and S. Then the second relation in (9.3) may be replaced by:

$$- k_1[E][S] + k_{-1}[ES] = 0 \qquad (9.5)$$

The formulae (9.4) and (9.5) give:

$$[ES] = [E]_0[S]/(k_{-1}/k_1 + [S]) \qquad (9.6)$$

and:

$$[\dot{P}] = k_2[E]_0[S]/(k_{-1}/k_1 + [S]) \qquad (9.7)$$

b) <u>Briggs and Haldane hypothesis</u>

This "quasi-stationary" hypothesis asserts that:

$$[\dot{ES}] \qquad (9.8)$$

may be neglected.

Thus the third relation in (9.3) may be replaced by:

$$k_1[E][S] - (k_{-1} + k_2)[ES] = 0 \qquad (9.9)$$

giving:

$$[\dot{P}] = k_2[E]_0[S]/((k_{-1} + k_2)/k_1 + [S]) \qquad (9.10)$$

We see that in both a) and b) the speed of transformation of S into P, S $\rightarrow$ P, is given by:

$$v = V_M[S]/(K_M + [S]) \qquad (9.11)$$

where V_M and K_M are the Michaelis constants.

Note that a reversible reaction such as the following is also possible:

$$E + S \underset{k_{-1}}{\overset{k_1}{\rightleftharpoons}} ES \underset{k_2}{\overset{k_{-2}}{\rightleftharpoons}} E + P \qquad (9.12)$$

The kinetic equations are easily obtained by a mass balance

and the Briggs-Haldane hypothesis gives:

$$v = [\dot{P}] = V_M[([S] - a [P])/(K_M + [S] + b[P])] \qquad (9.13)$$

where:

$$a = (k_{-1}/k_1)(k_{-2}/k_2) \quad , \quad b = k_{-2}/k_1 \qquad (9.14)$$

These properties and modelling allow the study of porous membranes separating two compartments. The enzymes are fixed on the membrane and the substrate is introduced at time $t = 0$ in the two compartments. The substrate diffuses through the membrane and the enzymes catalyse the transformation.

Let us now consider an irreversible mono-enzymatic system where the Dirichlet conditions are satisfied. Suppose that compartments I and II contain the substrate S and some co-factors which do not play a role. The substrate concentration is such that diffusion through the membrane does not change the concentration of S in either compartment. This concentration is asssumed to be constant and P is equal to zero. The speed of reaction for an elementary volume through the membrane is given by:

$$v = V_M[S]/(K_M + [S]) \qquad (9.15)$$

(mono-enzymatic and irreversible reaction).

For an elementary volume $\partial S/\partial t'$ global is due to two phenomena: reaction and diffusion.
The reaction part is:

$$[\partial[S]/\partial t']_{\text{enzymatic reaction}} = - V_M[S]/(K_M + [S])$$

The diffusion part (see Chapter 1) is equal to:

$$[\partial[S]/\partial t']_{\text{diffusion}} = D_S \partial^2[S]/\partial x'^2$$

where x' = distance (in cm) from the membrane point to the boundary between compartment I and the membrane, t' = time in hours ($t' > 0$), $[S]$ = concentration (in moles cm^{-3}) of substrate. $[S]$ is a function of x' and t'. D_S is the diffusion coefficient of S (in $cm^2 h^{-1}$), V_M the maximum speed (in moles $cm^{-3}h^{-1}$), K_M the Michaelis constant (moles cm^{-3}).

Some further parameters and variables now introduced:

ρ = membrane thickness, $[P]$ = concentration of P (moles cm^{-3}) and D_p = diffusion coefficient of P (cm^2h^{-1}).

The partial differential equation governing the variation of S in the membrane is therefore:

$$[\partial[S]/\partial t'] - D_S \partial^2[S]/\partial x'^2 + V_M[S]/(K_M + [S]) = 0 \quad (9.16)$$

Dimensionless variables may be introduced:

$$x = x'/\rho \qquad\qquad (0 < x < 1)$$

$$t = t'/(\rho^2/D_S) \qquad (0 < t < T) \qquad (9.17)$$

$$s = [S]/K_M \qquad\qquad P = [P]/K_M$$

to give the simplified equation:

$$\partial s/\partial t - \partial^2/\partial x^2 + \sigma s/(1+s) = 0 \qquad (9.18)$$

with $\sigma = (V_M/K_M)(\rho^2/D_S)$

The boundary conditions are:

$$s(0,t) = \alpha \quad , \quad s(1,t) = \beta \quad (\alpha \text{ and } \beta > 0) \qquad (9.19)$$

and the initial condition is:

$$s(x,0) = 0 \qquad (9.20)$$

To find p, the following equation may be obtained using the previous technique:

$$\partial p/\partial t - (D_p/D_S) \partial^2 p/\partial x^2 - \sigma s/(1+s) = 0 \qquad (9.21)$$

with the conditions:

$$p(0,t) = p(1,t) = 0, \quad p(x,o) = 0 \qquad (9.22)$$

We see that there is a great similarity with the modelling of respiration.

The difference in the present case is that enzymatic reactions introduce non-linearities in the partial differential equations. When considering different biological systems, other limit conditions are possible. A complete theoretical and numerical study is given in [48].

9.2 Numerical resolution of partial differential equations [56], [23]

9.2.1 SEMI-DISCRETIZATION TECHNIQUE

Consider a concrete example from the gas exchange problem (Chapter 1). The equations for oxygen are as follows:

$$\partial y/\partial t = K\partial^2 y/\partial x^2 - (1/S)V \partial y/\partial x - (2D/Re)(y - z)$$

$$\partial z/\partial t = -(Q/S_1)s_{O_2} \partial z/\partial x - (2\pi RD/eS_1)(z-y)$$

(9.23)

where we assumed V to be independent of x.

It is possible to reduce (9.23) to an ordinary differential system by setting:

$$y = \sum_{m=1}^{n} A_m(t) \exp(\beta_m x)$$

(9.24)

$$z = \sum_{m=1}^{n} B_m(t) \exp(\beta_m x)$$

where the β_m are either fixed or unknown. Here the β_m are assumed known but later we shall consider the identification of the β_m. Putting (9.24) into (9.23) and identifying the coefficients of $\exp(\beta_m x)$ gives the ordinary differential system:

$$\dot{A}_m = (K\beta_m^2 - (V(t)/S)\beta_m - (2D/Re))A_m(t) + (2D/Re)B_m(t)$$

$$\dot{B}_m = D(2\pi R/eS_1)A_m(t) - (s_{O_2}(Q/S_1)\beta_m + D(2\pi R/eS_1))B_m(t)$$

(9.25)

Other conditions (initial or final) are necessary to solve (9.25) when the paramters K, β_m and the function $V(t)$ are known. In this case the biological system allows initial conditions $(y(x,0),z(x,0))$ or boundary conditions such as $(y(0,t),z(0,t))$ to be given. For example, if $y(x,0)$ and $z(x,0)$ are given, we have:

$$y(x,0) = \sum A_m(0)\exp(\beta_m x)$$

$$z(x,0) = \sum B_m(0)\exp(\beta_m x)$$

$$(9.26)$$

that determine the initial conditions $A_m(0)$ and $B_m(0)$ using an optimization technique for calculating the minimum of J_1 and J_2 with:

$$J_1 = \int_0^L (y(x,0) - \sum A_m(0)\exp(\beta_m x))^2 dx$$

$$(9.27)$$

$$J_2 = \int_0^L (z(x,0) - \sum B_m(0)\exp(\beta_m x))^2 dx$$

Since J_1 and J_2 of (9.27) are second degree polynomials it is possible to find $x_m = A_m(0)$ and $y_m = B_m(0)$ by solving the linear algebraic systems:

$$\partial J_1/\partial x_m = 0 \quad , \quad \partial J_2/\partial y_m = 0 \quad , \quad m = 1, \ldots n \qquad (9.28)$$

Then the differential system (9.25) becomes solvable because the initial conditions are known [24]. The solution of the system (9.23) is therefore given by (9.24). This method is readily applicable when the partial differential equations are linear. When non-linearities arise, such as with the system:

$$\partial y/\partial t = K\partial^2 y/\partial x^2 -(1/S)\partial/\partial x(y.z) - (2D/Re)(y - z)$$

$$\partial z/\partial t = -(Q/S_1)s_{0_2} \partial z/\partial x - (2\pi RD/eS_1)(z-y)$$

$$(9.29)$$

the previous method may be used by linearizing the first equation with respect to x, on small intervals (x_i,x_{i+1}). On each interval (x_i,x_{i+1}), (9.29) gives a linear system and thus the semi-discretization technique just described can be used. The use of an optimization method is also possible as will be seen in the next section.

9.2.2 OPTIMIZATION METHOD FOR SOLVING PARTIAL DIFFERENTIAL EQUATIONS

Here more general approximations than in (9.26) are introduced.

$$y(x,t) = \sum_{i=1}^{N} a_i r_i(x) s_i(t)$$

$$z(x,t) = \sum_{i=1}^{N} b_i r_i(x) s_i(t)$$

(9.30)

where the $r_i(x)$, $s_i(t)$ are known functions such that the set:

$$\left\{ r_i(x) \; s_i(t) \right\}_{i=1}^{\infty}$$

may approach every "smooth" function of two variables (x,t) The previous functional set may be derived from a tensorial product [59] of functions of x and of functions of t only. Substituting approximations (9.30) into (9.23) and introducing a functional which integrates equations (9.23) between $(0,L)$ and $(0,T)$ leads to:

$$
\begin{aligned}
J = &\int_0^T \int_0^L \Big(\sum_{i=1}^{N} (a_i r_i \dot{s}_i - K a_i \ddot{r}_i s_i + (1/S) a_i \dot{r}_i s_i \\
&\quad + (2D/Re)(a_i - b_i) r_i s_i) \Big)^2 dxdt \\
+ &\int_0^T \int_0^L \Big(\sum_{i=1}^{N} ((\varrho/S_1) s_{O_2} b_i \ddot{r}_i s_i + b_i r_i \dot{s}_i \\
&\quad + (2\pi DR/eS_1)(b_i - a_i) r_i s_i) \Big)^2 dxdt
\end{aligned}
$$

(9.31)

The functional J depends only on the unknowns a_i and b_i, $(i = 1, \ldots, N)$. Its minimum can be found by solving the <u>linear</u> algebraic system:

$$\partial J/\partial a_i = \partial J/\partial b_i = 0 \quad , \quad i = 1, \ldots N \quad (9.32)$$

A true solution of (9.23) is obtained if $J = 0$ at the optimum. This method is possible because of the linearity of system (9.23). When considering a non-linear system such as (9.29) an optimization technique (local or global) is necessary. For non-linear integral equations (Chapter 8) some possible linearization methods still remain. Let us give an example:

$$\partial y/\partial t + ay^2 + by = f(x,t)$$

$$y(o,x) = y_0(x)$$

(9.33)

where a, b, $f(x,t)$ and $y_0(x)$ are fixed.

Taking into account the approximation (9.30) in (9.33) gives:

$$\sum_{i=1}^{N} (a_i r_i(x)\dot{s}_i(t) + a(\sum_{1}^{N} a_i r_i(x)s_i(t))^2 + b(\sum a_i r_i s_i)$$

$$= f(x,t) \qquad (9.34)$$

Writing (9.34) for some particular set (x_k, t_k), $k = 1, \ldots m$ gives a non-linear algebraic system:

$$\sum_{i=1}^{N} (a_i r_i(x_k)\dot{s}_i(t_k) + a(\sum_{i=1}^{N} X_i r_i^2(x_k)s_i^2(t_k)$$

$$+ 2a \sum_{\substack{i,j=1 \\ i<j, i\neq j}}^{N} X_{ij} r_i r_j s_i s_j + b(\sum a_i r_i(x_k)s_i(t_k))$$

$$= f(x_k, t_k) \qquad (9.35)$$

where we set $X_i = a_i^2$, $X_{ij} = a_i a_j$ $(i \neq j, i < j)$

When m is large enough that we have as many equations as unknowns in (9.35), a linear system is obtained. Its solution gives an approximate solution of (9.33). This technique leads to the solution when (9.35) is a non-degenerate system. In other words this means that the set of functions:

$$\left\{ r_i \dot{s}_i, \ r_i^2 s_i^2, \ r_i s_i r_j s_j, \ r_i s_i \right\}$$

is linearly independent.

Another method using an idea developed from non-linear integral equations and based on the regularity constraint (Chapter 8) can be proposed. The solution will be sought using the formula:

$$y(x,t) = \sum_{i=1}^{N} r_i(x)s_i(t) \qquad (9.36)$$

where the functions $s_i(t)$ are fixed. Only the $r_i(x)$ have to be identified.

Integrating (9.33) with respect to t gives:

$$\sum r_i(x) \int_0^T \dot{s}_i(t)dt + a \int_0^T (\sum r_i(x)s_i(t))^2 dt$$

$$+ b \sum r_i(x) \int_0^T s_i(t)dt = \int_0^T f(x,t)dt \qquad (9.37)$$

Making $x = a_j$, then $x = b_j$ such that $b_i < a_i < b_{i+1}$ in (9.37) gives an algebraic system. The unknowns chosen are $r_i(a_j)$ and $r_i(a_k)r_j(a_k)$. As in Chapter 8, a regularity constraint such as the following is chosen:

$$r_i(b_k) = \sum_{l=0}^{n} r_i(a_l)T_l^{(i)}(b_k) \qquad (9.38)$$

where the polynomials $T_l^{(i)}$ are well-defined.

The algebraic system becomes linear with respect to the unknowns $r_i(a_j)$ and $r_i(a_k)r_j(a_k)$. It is sufficient to choose n such that the a_k, b_k, $k = 1, . . . ,n$ lead to a linear algebraic system with at least as many equations as unknowns. The linear relations:

$$y_0(a_j) = \sum_{j=1}^{N} r_i(a_j)s_i(0) \qquad (9.39)$$

have be added to the previous one to take the initial condition into account.

9.2.3 SOLUTION OF PARTIAL DIFFERENTIAL EQUATIONS USING A COMPLETE DISCRETIZATION [56], [57]

We return to the differential systems (9.25). They can be discretised using a numerical method such as the Euler technique [24]. A functional system such as the following is obtained:

$$(A_m(t_{i+1}) - A_m(t_i))/(t_{i+1} - t_i) = F_{1m}(\gamma(t_i), A_m(t_i), B_m(t_i))$$

$$(B_m(t_{i+1}) - B_m(t_i))/(t_{i+1} - t_i) = F_{2m}(Q(t_i), A_m(t_i), B_m(t_i))$$

with $i = 0, . . ,N;$ $t_i \in [0,T]$, $m = 1, . . ,n$.   (9.40)

F_{1m}, F_{2m} are functions that may be found from the known parameters of (9.23). If the initial conditions $A_m(0)$, $B_m(0)$ are known or calculated the system (9.40) may be explicitly solved. We obtain successively $A_m(t_1)$, $B_m(t_1)$, $A_m(t_2)$, $B_m(t_2)$ and so on.

Another complete discretization is obtained by directly approximating the partial derivatives in the system (9.23). For example in the first equation (9.23) we obtain a discretization in time $t = rk$ (k = discretization step) and in space $x = nh$ ($h > 0$ is the discretization step). Using the following approximate formulae:

$$\partial y/\partial t(nh,rk) = (y(nh,(r+1)k) - y(nh,rk))/k$$

$$\partial^2 y/\partial x^2(nh,rk) = (y((n+1)h,rk - 2y(nh,rk) + y((n-1)h,rk))/h^2$$

$$\partial y/\partial x(nh,rk) = (y((n+1)h,rk) - y(nh,rk))/h   (9.41)$$

Putting these expressions in the first equation (9.23) gives the discretised relation:

$$y_n^{r+1} - y_n^r = k(K(y_{n+1}^r - 2y_n^r + y_{n-1}^r)/h^2$$

$$- (1/S)\gamma^r(y_{n+1}^r - y_n^r)/h - (2D/Re)(y_n^r - z_n^r)   (9.42)$$

with $r = 0, 1, . . $; $n = 1, 2, . . $ and where we set $f(nh,rk) = f_n^r$.

A similar result is obtained for the second relation in (9.23). The unknowns are now y_n^r and z_n^r (if γ is known). As can easily be seen, $y(x,0)$ and $z(x,0)$ are given or, if we prefer, when y_n^0 and z_n^0 are given for all n, we can calculate successively y_n^1, z_n^1 and y_n^2, z_n^2 and so on. Results about the convergence of such procedures can be found in [56] or [57].

<u>Remark</u> Other methods such as the Alienor transformations can be used for solving partial differential equations [15]. We obtain functional relations with one variable.

9.3 Identification in partial differential equations

As in compartmental analysis where some parameters in differential systems have to be identified, the partial differential equations can contain some unknown parameters. The identification is easy when these parameters occur linearly. Consider for example the following equation:

$$\partial y/\partial t = a\partial^2 y/\partial x^2 + b\partial y/\partial x + c \qquad (9.43)$$

where the coefficients a ,b ,c have to be identified from the knowledge of y(t), $t \in [0,\infty]$.
    A functional J is introduced:

$$J = \int_0^T \int_0^L (\partial y/\partial t - a\partial^2 y/\partial x^2 - b\partial y/\partial x - c)^2 dt dx \qquad (9.44)$$

which depends only on a, b and c. Since J is a second-degree polynomial the optimum (minimum) is obtained by writing the relations:

$$\partial J/\partial a = \partial J/\partial b = \partial J/\partial c = 0 \qquad (9.45)$$

The system (9.45) is linear and its solution gives a*, b*, c*. This solution is applicable if $J(a^*,b^*,c^*) = 0$.
    In general, we shall have only $y(x_j,t_j)$ for j = 1, . . ,m. In this case an approximation can be chosen for y:

$$y(x,t) = \sum_{i=1}^{N} a_i r_i(x) s_i(t) \qquad (9.46)$$

where a_i, r_i and s_i are known (possibly by using an optimization technique).

Using (9.43), (9.46) and a new functional J_1 we obtain:

$$J_1 = \sum_{i=1}^{m} [\sum a_i r_i(x_j)\dot{s}_i(t_j) - a \sum a_i \ddot{r}_i(x_j)s_i(t_j)$$

$$- b \sum a_i \dot{r}_i(x_j)s_i(t_j) - c]^2$$

(9.47)

whose minimum is obtained by writing:

$$\partial J_1/\partial a = \partial J_1/\partial b = \partial J_1/\partial c = 0 \qquad (9.48)$$

The solution of the linear algebraic system (9.48) gives the solution of our identification problem in the "discrete" case. When non-linearities arise in the partial differential equations the functionals to minimise can be written using the previous technique. The optimization problems will be reduced, as before, to the solution of a linear system if the parameters to be identified occur in a linear manner. If they occur non-linearly, optimization techniques become necessary for their determination. A method for determining some coefficients by the global optimization technique (Alienor) will be presented in the following.

9.4 Optimal control with partial differential equations

Consider the respiration problem modelled in Chapter 1. The partial differential equations are paired. In fact, we have two different relations for oxygen, and so on. Let us write the two equations relating to carbon dioxide (CO_2). Now a simpler notation is used, but the precise relations may be found in Chapter 1.

$$v'_t = K_a v''_{x^2} - g(t)v'_x - d_1(v - w)$$

$$w'_t = - sqw'_x + d_2(v - w)$$

(9.49)

where $g(t)$ is the convection flow ($\gamma(t)$ in our previous notation), K_a, d_1, and d_2 are diffusion coefficients and q is the blood flow assumed constant so as to simplify our optimization problem.

Classical literature [75] gives values of physiological parameters and we can write (9.49) in a more explicit form:

$$v'_t = 11v''_{x^2} - 49g(t)/49.v'_x - 163.35(v - w)$$

$$w'_t = -235.2w'_x + 1453.6(v - w)$$

(9.50)

It is not difficult to verify that (9.50) is a stiff partial differential equation. The classical methods [56] fail when trying to solve (9.50) numerically. Furthermore, a classical method cannot be used with a mini-computer because too many calculations are required. Indeed, when considering a parabolic equation [56]:

$$u'_t - au''_{x^2} - bu'_x + ug = f(x,t) \qquad (9.51)$$

for $t \in [0,T]$ $\quad |a(x,t)| \leqslant A$, $|b| \leqslant MA$, $|g| \leqslant G$

it is necessary to satisfy the following inequalities for the discretization steps:

$$h < 2/M \text{ and } \tau < h^2/(2A + Gh^2) \qquad (9.52)$$

(h = step in space, τ = time)

so as to ensure the convergence of the discretization method.

In our case we find:

$$A = 11, \quad M = 49/11 = 4.46, \quad G = 163.35 \qquad (9.53)$$

and therefore:

$\quad$ h < 0.448, which implies 656 steps in space $\qquad$ (9.54)

and:

$$\tau < 0.448^2/(22 + 163.35 \times 0.448^2) = 3.66 . 10^{-3} \qquad (9.55)$$

implying 391 steps in time.

In total we need 391 x 656 = 256469 points of R^2. This number exceeds the possibilities of microcomputers. Furthermore these discretization techniques require knowledge of $v(x,0)$ and $w(x,0)$. In our system they are not known and thus a different technique is required. The method using characteristic functions [23] will be used. Approximations of v and w are useful for it.

$$v(x,t) = \sum_i \exp(\lambda_i x)\phi_i(t)$$

$$w(x,t) = \sum_i \exp(\lambda_i x)\psi_i(t) \qquad (9.56)$$

Substituting these in (9.49) gives the following ordinary differential system:

$$\dot{\phi}_i = (K_a\lambda_i^2 - \lambda_i g(t) - d_1)\phi_i + d_1\psi_i$$

$$\dot{\psi}_i = d_2\phi_i - (sq\lambda_i + d_2)\psi_i \qquad (9.57)$$

Unfortunately the $\phi_i(0)$ and $\psi_i(0)$ are unknown. But these equations (9.57) are homogeneous, so we can assume:

$$\phi_i(0) = \psi_i(0) = 1 \quad \text{in the solution.} \qquad (9.58)$$

In (9.56) at least three λ_i are necessary and we can set:

$$v(x,t) = \sum_{i=0}^{2} \exp(\lambda_i x)\phi_i(t)$$

$$\phi_i = a_{i0}(1 + a_{i1}t + a_{i2}t^2 + a_{i3}t^3)$$

$$w(x,t) = \sum_{i=0}^{2} \exp(\lambda_i x)\psi_i(t) \qquad (9.59)$$

$$\psi_i = b_{i0}(1 + b_{i1}t + b_{i2}t^2 + b_{i3}t^3)$$

by choosing a polynomial approximation for ϕ_i and ψ_i. The constant terms are necessary in ϕ_i and ψ_i because of the homogeneity of (9.57).

The resolvents of the system (9.57) are now introduced. They are obtained by eliminating ψ_i of the first equation in (9.57) and ϕ_i of the second. We have:

$$\ddot{\phi}_i + (d_1 + d_2 + (g(t) + sq - \lambda_i K_a)\lambda_i)\dot{\phi}_i$$

$$+ (g' + sqd_1 + (g - \lambda_i K_a)(sq\lambda_i + d_2)).\lambda_i\phi_i = 0 \qquad (9.60)$$

$$\ddot{\psi}_i + (d_1 + d_2 + (g(t) + sq - \lambda_i K_a)\lambda_i)\dot{\psi}_i$$

$$+ (sqd_1 + (g(t) - \lambda_i K_a)(sq\lambda_i + d_2)).\lambda_i\psi_i = 0$$

Three particular points t_i from Gauss quadrature are used:

$$t_0 = 0.1127\ T,\quad t_1 = T/2,\quad t_3 = 0.8873\ T \qquad (9.61)$$

where T is the time of inspiration.
The numerical algorithm is as follows:

(i) First suppose that the λ_i (i = 0, 1, 2) are determined then the use of equation (9.60) (with the relations (9.59)) written at t_0, t_1, t_3 gives six linear algebraic relations to identify a_{i1}, a_{i2}, a_{i3}, b_{i1}, b_{i2}, b_{i3} for each i.

(ii) The identification of the a_{i0} and b_{i0} (i = 0, 1, 2) remains. Three physiological relations are available for the a_{i0}:

$$1/T \int_0^T v(0,t)dt\ =\ 34, \qquad 1/T \int_0^T v(L,t)dt\ =\ 40,$$

$$(1/T.L) \int_0^L dx \int_0^T v(x,t)dt = 38 \qquad (9.62)$$

where L is the length of the tube (see Chapter 1). Three other relations are available for the b_{i0}. They are:

$$1/T \int_0^T w(0,t)dt\ =\ 46, \qquad 1/T \int_0^T w(L,t)dt\ =\ 40,$$

$$(1/T.L) \int_0^L dx \int_0^T w(x,t)dt = 42 \qquad (9.63)$$

These relations, linear in the a_{i0} and b_{i0}, give the solution.

(iii) Now the λ_i (i = 0, 1, 2) have to be identified. They enter in a non-linear way, so a global optimization method is indicated. In equations (9.50) and (9.57) only the ventilation function g(t) is given; the blood flow q is unknown and has to be identified. Without a supplementary criterion it would not be possible to identify the four parameters appearing in (9.50) or (9.57). For this case A. Guillez proposed a mechanical criterion based on the following result: the amount of CO_2 lost by the blood must be equal to that gained by the alveolus (tube). The criterion can be expressed as follows [15]:

$$DE = \left| \; 1 - \int_0^T Sg(t)(v(L,t) - v(0,t))dt \; / \; 6S_1qT \; \right| . \; 100 \qquad (9.64)$$

This has to be minimised as a function of q and the λ_i. We see that q appears explicitly in the expression for DE, and the λ_i are present in:

$$v(L,t) = \sum_{i=0}^{2} \exp(\lambda_i.L)\phi_i(t)$$

To obtain an optimal solution minimising DE, that is:

$$\text{Min} \quad DE(q, \; \lambda_i, \; a_{ij}, \; b_{ij}) \qquad (9.65)$$

the following iterative process may be used:

a)   By the Alienor transformation the variables q and λ_i are reduced to a single variable θ. Once the variable θ has been fixed the a_{ij}, b_{ij} are determined by previously described methods.

b)   We look for a minimum of DE as a function of θ by iteration.   In practice a change of θ involves a new calculation of a_{ij}, b_{ij} and therefore a new value of DE. Once θ is fixed therefore, DE may be calculated because the values of q and λ_i depend on θ, and the values of a_{ij} and b_{ij} can also be obtained.   It is clear that this method gives a global optimum of DE.   Indeed a global minimization in terms of θ implies a global minimization in terms of and a_{ij} and b_{ij}, since the latter are functions of θ.

Remark 1 The function g(t) chosen by A. Guillez is:

$$g(t) = 5.925 \; t^3 - 50.83 \; t^2 + 109 \; t \quad (\text{in } \mu m/s) \qquad (9.66)$$

The partial pressures of gas are expressed in torrs.

Remark 2 The foregoing demonstration is based on the following property:

If we want to minimise f(x,y) and if y can be expressed

explicitly as a function of x by y = g(x) then the minimization of f(x,y) is equivalent to minimising f(x,g(x)) = h(x) which depends only on x.

Some new difficulties arise when considering non-linear equations. Formulae (9.56) can still be used, but differential systems such as (9.57) may not be written. Methods developed in section 9.2 have to be used, or simpler models must be introduced. An important fact concerning the foregoing technique using Alienor should be noted. The blood flow q is determined for each ventilation function g(t). In practice this relation is unknown and one of the main interests of the numerical technique is to find it. More precisely, g(t) depends on the frequency and amplitude of respiration. Thus the relation between frequency, amplitude, and blood flow q can be identified. In fact, the respiratory system is self-governing. It is sufficient to determine the ventilation g(t) to give the blood flow q. Furthermore, numerical results show that a concentration balance is obtained at the end of the tube (x = L) when inspiration is finished.

9.5 Other approaches for optimal control

We use the previous approximations:

$$v(x,t) = \sum A_m(t) \, \exp(\beta_m x)$$
$$w(x,t) = \sum B_m(t) \, \exp(\beta_m x) \tag{9.67}$$

where the summations are finite.

We obtain the differential system (9.40) which may be solved and gives the $A_m(t_i)$, $B_m(t_i)$ (by a numerical method) at different times t_i as functions of $V_i = V(t_i)$ and $Q_i = Q(t_i)$. In this approach V and Q are considered as functions of t. A control problem may thus be introduced in connection with a physiological property. In fact it can be seen that $W(L,t)$ is a regulated value which is always close to 0.05 bar. The following optimal control problem results:

$$\underset{V_i, Q_i}{\text{Min}} \int_0^T (W(L,t) - 0.05)^2 dt \simeq \underset{V_i, Q_i}{\text{Min}} \sum_{i=0}^N d_i(W(L,t_i) - 0.05)^2 \tag{9.68}$$

by using a quadrature formula. The d_i are known and come
from the quadrature; T is the repiratory period. From the
solution of (9.40), W(L,t) is a function of γ_i and Q_i and
thus the problem (9.68) is well expressed. In fact $W(L,t_i)$
is a rational function of γ_i, Q_i and (9.68) has at least one
minimum. If uniqueness is not assured, physiologists may
give some complementary relations, such as:

$$\gamma(t) = F(Q(t)) \qquad\qquad (9.69)$$

giving a unique solution for the control problem (9.68).
This approach is slightly different to the previous one and
may need more powerful computers.

Another possibility which does not seem to have been
tried in this context is to look for a functional relation
such as the following:

$$W(L,t) = F(Q(t), \gamma(t)) \qquad\qquad (9.70)$$

where the forms of Q(t) and $\gamma(t)$ are known (for instance by
using the previous studies). The identification of the
function F from experimental data available in the classical
literature remains. The form of F may be chosen, a priori,
such that only some parameters have to be identified. Then
(9.68) becomes a classical optimization problem and a local
or global optimization technique can be employed. Once the
optimal functions γ^* and Q^* have been calculated the
solution of the partial differential system in v and w gives
the optimal states v*(t) and w*(t).

9.6 Other partial differential equations

Modelling a single element of the kidney leads to partial
differential equations. Classical theories give equations
similar to the repiratory equations. Some contradictions
arise with the physiological data. A. Guillez [31] proposed
a novel approach based on a directivity hypothesis. Active
and passive transport [31] cannot explain the phenomenon of
material exchanges between membranes. The main substances
playing a part in the working of the kidney are salt, urea,
and water. The Guillez hypothesis takes the orientation of
the diffusion phenomena into account. It seems to be
necessary and sufficient [31] to explain the electrolyte
balance in the human organism. The contradictions

associated with classical theories disappear. Taking a tube
(as for respiration) of length L and cross-section σ, let us
represent the concentration of a substance in the tube by
C_i, the exterior concentration (in the interstitial space)
by C_e and the concentration in the wall by C. For passive
transport the following equation is obtained:

$$\partial C_i / \partial t + \sigma^{-1} \partial / \partial x (v C_i) = a\sigma^{-1} D(C_e - C_i) + D_x \partial^2 C_i / \partial x^2$$
(9.71)

taking into account diffusion, convection and diffusion
through the wall.

In (9.71), v represents the flow of the substance and a
is the radius of the tube. D_x is the longitudinal diffusion
coefficient and D the coefficient from Fick's Law. A mass
balance in a small element gives:

$$- \partial v / \partial x = K(p(x) - p_e) + k(C_e - C_i)$$
(9.72)

where the coefficients K and k are the parameters of wall
permeability and osmosis. p(x) is the hydrodynamic pressure
and p_e the pressure of the substance in the tissues. (9.71)
ignores active transport. In the last case equations such
as the following are obtained:

$$\partial C_i / \partial t + \sigma^{-1} \partial / \partial x (v C_i) = - a\sigma^{-1} D C_i + D_x \partial^2 C_i / \partial x^2$$
(9.73)

Now let us consider directive diffusion and osmosis. The
expression $(C_e - C)$ corresponds to the transversal
coefficient D_{ie} from the outside to the inside, and the
coefficient of osmosis k_{ie}. Similarly, k_{ei} and D_{ei} are
associated with the difference $C_i - C$. These coefficients
are, in general, different. Equations (9.71) and (9.72)
become:

$$\partial C_i / \partial t + \sigma^{-1} \partial / \partial x (v C_i) = a\sigma^{-1} (D_{ie} C_e - D_{ei} C_i) + D_x \partial^2 C_i / \partial x^2$$
(9.74)

which describes a "directive" transport.

$$- \partial v / \partial x = K(p(x) - p_e) + k_{ie} C_e - k_{ei} C_i$$
(9.75)

represents a "directive" osmosis.

The study of these equations [31] allows us to propose
a theory of kidney function. It is consistent with all

known experimental data. Sometimes imagination is necessary for solving difficult problems. Analogies with physical systems (transistors in this case) are often useful. Of course it is not difficult to propose optimal control problems related to kidney function. For example, a problem could be: How can a constant concentration $C_i(L,t)$ be maintained as closely as possible?

CHAPTER 10

OPTIMALITY IN HUMAN PHYSIOLOGY

10.1 General remarks

Many biological or physiological systems cannot be studied
without the introduction of supplementary relations. The
equations obtained from a physical and chemical study are
often insufficient [7] to obtain a mathematically well-
defined problem. In general, the uniqueness of solutions is
not assured, and so it is necessary to find some other
relations or criteria. In practice, an optimality criterion
is often chosen. The most intuitive one is to introduce the
following idea: the system will use minimal energy to
attain some biological function.

This basic idea has often been used in physics and
biology. But while it was quantified in physics, it often
remained unquantified in biology and medicine. It is only
recently that optimality criteria have been introduced into
mathematical modelling associated with biological phenomena.
In the following some concrete examples from biomedicine
will be treated. The models were designed following the
suggestions of the French physiologist Jean Brocas
(University Paris 5). It is always difficult to justify the
introduction of a criterion to be optimised such as, for
example, the minimization of the energy consumed by a
biological system. We can only say that such a criterion
allowed us to account for and explain the experimental data
in the examples that we considered. We suggest that other
optimization criteria could be useful because we know that
approximations (of functions, models, . .) are never unique.
Nevertheless if a model using an optimization criterion is
consistent with the experimental data, and the biophysical
properties of the system being studied, it can be used for
prediction and action can be taken until contradictions
arise.

10.2 A mathematical model for thermo-regulation [30]

In Kayser [43] the heat transport system of the human body is shown, in a diagrammatic form, as follows:

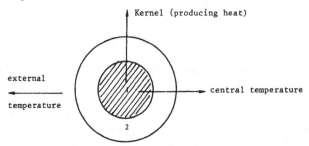

The exchanges between 1 and 2 are the core-boundary transport, and those between 2 and the exterior are the boundary-ambient atmosphere transport.

The following principle will be used for modelling the heat exchanges in the human body.

When there is a temperature difference between two areas, the heat goes from the warm area to the cold area, such that the temperature difference tends towards zero. These heat transfers result from four mechanisms:

a) conduction, where contact between two bodies ensures heat exchange;

b) convection, which is in some ways similar. It allows the exchange of heat between a fluid and a solid body. It involves conduction at the fluid-solid boundary and a mass transfer in the fluid;

c) radiation, which gives heat exchange through electromagnetic waves;

d) evaporation, which is a change from the liquid to the gaseous phase involving transfer of heat.

A first equation [30] may be written for the transfer of heat through the continuous medium of the tissues (conduction and convection phenomena are involved).

$$Q = k(T_1 - T_2) \qquad (10.1)$$

where k is a known parameter.

A second equation represents the heat transport from the boundary to the exterior. We can write:

$$Q_{conv} = \alpha S(T_s - T_a) \tag{10.2}$$

where T_s is the skin temperature and T_a the ambient temperature.

A third equation represents the radiation and involves the area of the body. We can write:

$$Q_{rad} = \sigma S(T_s^4 - T_w^4) \tag{10.3}$$

where T_s and T_w are the respective temperatures.

A final equation is obtained by considering water evaporation at the boundary:

$$Q_{evap} = kS(P_s - P_a) \tag{10.4}$$

where S is the skin surface, k a parameter, P_s is the output pressure and P_a the partial pressure of water vapour in the atmosphere.

The following diagram summarises the heat exchanges in the body:

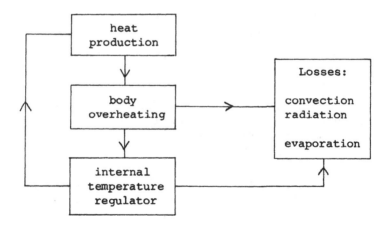

The peripheral system of thermal loss is represented in Figure 14.

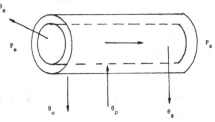

FIG. 14

The following notation will be used:

θ_e = liquid temperature at entrance of tube

θ_s = liquid temperature at exit of tube

θ_p = envelope temperature

θ_0 = external temperature

P_e = pressure at entrance of tube

P_s = pressure at exit of tube

q = liquid flow in the tube

P_w = partial pressure of steam on the boundary

P_a = partial pressure of steam in the atmosphere

N = specific heat of the liquid

D = diffusion coefficient of the tube wall

h_e = evaporation coefficient

h_c = convection parameter

A = envelope surface

The aim of the model is to study the heat exchanges between the interior and exterior of the cylinder as a function of the external temperature variations θ_0.

Three heat flows have to be considered:

(i)   The hot liquid in the tube has a heat flow similar to convection in the blood. This flow J_1 is equal to:

$$J_1 = q \cdot \Delta t \cdot N$$

where t is the temperature variation in the tube. With our notation:

$$J_1 = q(\theta_e - \theta_s)N \qquad (10.5)$$

J_1 must be positive and thus $\theta_s < \theta_e$

(ii) The diffusion phenomenon due to heat exchange between the interior of the tube and the wall gives a flow J_2:

$$J_2 = D.\Delta T' \qquad (10.6)$$

where $\Delta T'$ is the temperature variation.

We have $\Delta T' = (\bar{\theta}_s - \theta_p)$ with $\bar{\theta}_s$ = mean temperature in the tube. Taking the arithmetic mean between θ_e and θ_s for $\bar{\theta}_s$, gives:

$$J_2 = D((\theta_e + \theta_s)/2 - \theta_p) \qquad (10.7)$$

(iii) The heat exchange between the wall and the exterior involves conduction, convection, radiation and evaporation [30].

Neglecting the conduction and radiation gives:

$$J_{3conv} = - h_c(\theta_0 - \theta_p)A \qquad (10.8)$$

for the flow due to convection , and:

$$J_{3evap} = h_e(P_w - P_a)A \qquad (10.9)$$

for the evaporation flow.

Finally, we obtain:

$$J_3 = J_{3conv} + J_{3evap} = - h_c(\theta_0 - \theta_p)A + h_e(P_w - P_a)A \qquad (10.10)$$

Here we choose $P_a = 0$ giving:

$$J_3 = -h_c(\theta_0 - \theta_p)A + h_e P_w A \qquad (10.11)$$

For the steady state, we can write:

$$J_1 = J_2 = J_3 \qquad (10.12)$$

This balance is obtained from the following hypothesis:

The calorific contribution in the tube is <u>minimised</u>.

This gives the power minimization, that is:

$$\mathcal{J} = q \Delta p \qquad (10.13)$$

where Δp is the pressure difference in the tube which has to be minimised. With the notation of the model, we can write:

$$\mathcal{J} = q(P_e - P_s)$$

To summarise our results, let us write the mathematical formulation for our optimal system:

$$- h_c(\Theta_0 - \Theta_p)A + h_e P_w A = q(\Theta_e - \Theta_s)N$$

$$q(\Theta_e - \Theta_s)N = D(\bar{\Theta}_s - \Theta_p) \qquad (10.14)$$

$$q(P_e - P_s) \text{ is minimum}$$

Let us notice that $\mathcal{J} = q(P_e - P_s)$ minimum gives a first solution $P_e - P_s = 0$ which is not admissible. Therefore a constraint must be added:

$$P_e - P_s \text{ is assumed constant} \qquad (10.15)$$

The minimization of $\mathcal{J}$ is equivalent to that of q and the final model may be written as:

$$- h_c(\Theta_0 - \Theta_p)A + h_e P_w A = q(\Theta_e - \Theta_s)N$$

$$q(\Theta_e - \Theta_s)N = D((\Theta_e + \Theta_s)/2 - \Theta_p)$$

$$P_w = a\Theta_p + b \qquad (10.16)$$

$$q \text{ is minimum}$$

where the third equation is obtained on the assumption that P_w is an affine function of the membrane temperature. Furthermore it is an increasing function [30] and thus $a > 0$

The known parameters are:

$$h_c, h_e, A, N, \Theta_e, a, b$$

Θ_0 will be chosen such that $\Theta_0 \in [0,T]$ and $T < \Theta_e$ $\qquad$ (10.17)

   Only positive exterior temperatures will be considered.
The unknown parameters are:

$$\Theta_s, \Theta_p, P_w, q \qquad (10.18)$$

The variable D will now be discussed. It is in some sense
an unknown. If we suppose D = constant, then the optimal
solution is given by:

$$\Theta_s = \Theta_e = \Theta_p \qquad \text{giving } q = 0 \qquad (10.19)$$

D may thus be considered as a parameter function of Θ_0. A
hypothesis will be made about the structure of D such that
$D = D(\Theta_0)$.   Another mathematical possibility would be to
assume D to be an independent variable but in that case, the
uniqueness of the solution of (10.16) cannot be assured.   It
is easy to eliminate some parameters from equation (10.16)
so as to obtain a unique solution.

$$q = D(\Theta_0)[(\Theta_e + \Theta_s)/2 - (h_c\Theta_0 - bh_e)/(h_c + ah_e)]$$

$$/ [(\Theta_e - \Theta_s)N(1 + D(\Theta_0)/A(h_c + ah_e))]$$

$$(10.20)$$

where $D(\Theta_0)$, $\Theta_0 \in [0,T]$, is chosen as an increasing
function going to infinity for a critical value of Θ_0. This
is the value that releases the phenomenon of perspiration
(sudation).   The following relation is proposed:

$$D(\Theta_0) = c \exp(-\lambda/(\Theta_0 - \Theta_{su})) , \lambda > 0 \qquad (10.21)$$

where c and λ have to be determined. In (10.21) Θ_{su}
represents the temperature releasing the perspiration.
   Because of the monotonicity of Θ_0 the mathematical
problem is to minimise J defined by:

$$J(a_s,b_s,\lambda,\lambda_s,c) = 1/T \int_0^T q(a_s,b_s,\lambda,\lambda_s,c; \Theta_0)d\Theta_0 \qquad (10.22)$$

as a function of $a_s,b_s,\lambda,\lambda_s,c \in \mathbb{R}^5$ $(\lambda > 0)$

   The number of parameters to be identified can be
reduced using mathematical and biological relations.
   $\Theta_s(\Theta_0)$ may be assigned as:

$$\Theta_s(\Theta_0) = a_s \text{ th } \lambda_s\Theta_0 + b_s \qquad (10.23)$$

with a_s, λ_s, b_s to be identified, and th is the hyperbolic tangent.

(10.23) is obtained supposing that $\Theta_s(\Theta_0)$ is an increasing function of Θ_0. Furthermore, $\Theta_e - \Theta_s$ has to be bounded:

$$\alpha_1 \leqslant \Theta_e - \Theta_s \leqslant \alpha_2 \qquad (10.24)$$

From $0 \leqslant \text{th } \lambda_s\Theta_0 \leqslant 1$ and $\Theta_0 = 0$ we have $b_s = \Theta_e - \alpha_2$

At $\Theta_0 = T$ we obtain $0 \leqslant \text{th } \lambda_s T = (\alpha_2 - \alpha_1)/a_s \leqslant 1$ and thus:

$$a_s \geqslant \alpha_2 - \alpha_1$$

and λ_s may be expressed as a function of a_s:

$$\lambda_s = 1/2T \text{ Log } |(a_s + (\alpha_2 - \alpha_1)) / (a_s - (\alpha_2 - \alpha_1))| \qquad (10.25)$$

Furthermore it is now possible to calculate th $\lambda_s\Theta_0$:

$$\text{th } \lambda_s\Theta_0 = \frac{[(a_s + (\alpha_2-\alpha_1))/(a_s - (\alpha_2-\alpha_1))]^{\Theta_0/T} - 1}{[(a_s + (\alpha_2-\alpha_1))/(a_s - (\alpha_2-\alpha_1))]^{\Theta_0/T} + 1} \qquad (10.26)$$

The optimization problem (10.22) is reduced to:

Find $(a_s^*, \lambda^*) \in K_1$ minimising

$$J(a_s, \lambda) = 1/T \int_0^T q(a_s, \lambda; \Theta_0)d\Theta_0 \qquad (10.27)$$

with $K_1 = \left\{(a_s, \lambda) \quad R_+^2 \text{ such that } a_s > \alpha_2 - \alpha_1\right\}$

A first classical method is to find (a_s^*, λ^*) by solving the system:

$$\partial J/\partial a_s = \partial J/\partial \lambda = 0 \qquad (10.28)$$

In fact, q may be calculated explicitly as a function of a_s, λ, Θ_0. We have:

$$q = \frac{c\ \exp(-\lambda/(\Theta_0-\Theta_{su}))}{N(1 + [c\ \exp(-\lambda/(\Theta_0-\Theta_{su}))\ /\ A(h_c+ah_e)])} \times$$

$$\frac{[0.5\ (\Theta_e+\Theta_s) - (h_c\Theta_0-h_eb)/(h_c+ah_e)]}{(\Theta_e-\Theta_s)}$$

(10.29)

with $\Theta_s(\Theta_0) = a_s\ \text{th}\ \lambda_s\Theta_0 + b_s$ and

$$\lambda_s = 1/2T\ \text{Log}((a_s + (\alpha_2-\alpha_1))\ /\ (a_s - (\alpha_2-\alpha_1))$$

In [30] L. Dorveaux proved that it is possible to obtain:

$$\partial J/\partial a_s(a_s^*,.) = 0$$

but we cannot satisfy $\partial J/\partial \lambda\ \ (.,\lambda) = 0$

In consequence, an optimization technique is needed to find the minimum of J. A penalty function can be used to eliminate the equality or inequality constraints. The following optimization problem may be introduced:

Find $(a_s^*,\lambda^*) \in K_1$ such that

$$J(a_s^*,\lambda^*) = \text{Min}\ J(a_s,\lambda)$$

(10.30)

$$\partial J/\partial a_s(a_s^*,\lambda^*) = 0$$

with $K_1 = \left\{(a_s,\lambda) \in R^2 \text{ such that } a_s > \alpha_2 - \alpha_1 > 0,\ \lambda > 0\right\}$

The global optimization technique (Alienor) can be used to find the global minimum.

A more complicated model may be obtained by introducing the perspiration phenomenon. For more details [30] may be consulted. A numerical study used the following data:

$\Theta_0 \in [0,35°C]$,   T $= 35°C$, $\alpha_1 = 0.1$, $\alpha_2 = 2$, giving

$35 \leqslant \Theta_s \leqslant 36.9$, $\Theta_0 = \Theta_{su} = 35.5°C$, $p_W(\Theta_p) = 1.3\Theta_p - 8.3$

where p_W is in Torrs. Furthermore:

$N = 4180$ J/kg,   A $= 1.7$ m^2, $\Theta_0 = 28°C$, $\Theta_p = 33°C$,

p_W = 37.79 torrs, q = $1.02.10^{-5}$ m^3/s (approx. 10% of the cardiac flow), $\Theta_e - \Theta_s \simeq 0.7$. These data lead to:

h_c = $2.147.10^{-3}$ W/°C, h_e = $2.147.10^{-4}$ W/mm Hg, C = 10^{-3}.

The optimization method Alienor gives:

$$a_s^* = 765.748, \quad = 0.01077 \qquad (10.31)$$

Remark For practical minimization the functional used was:

$$\sum_{k=1}^{M} (y_k - q_k(a_s, \lambda))^2$$

where the y_k are experimental data. The previous constraints are added, especially $\partial J/\partial a_s(a_s, \lambda) = 0$. This is needed because we have the result: J does not have a minimum as a function of λ.

10.3 Optimization of pulmonary mechanics

We know the pulmonary system ensures gas exchanges between the alveolus and the atmosphere. More precisely, oxygen ($\dot{V}_{O_2}$) is brought to the tissues and carbonic gas is evacuated from the body ($\dot{V}_{CO_2}$). The gaseous flow varies according to the energy requirement of the organism. This gaseous transfer system [47] involves:

(i) a conduction, or dead, space, V_D;

(ii) an exchange zone or alveolar volume V_A in the blood of the capillaries of the lung.

The flow $\dot{V}$ is the product of V_T, tidal volume for each respiratory cycle ($V_T = V_D + V_A$) and the ventilation frequency, ν:

$$\dot{V} = V_T \nu = (V_D + V_A)\nu \qquad (10.32)$$

An infinite number of combinations (V_T, ν) can give the same flow $\dot{V}$. To each (V_T, ν) there corresponds an alveolar ventilation:

$$\dot{V}_A = V_A \cdot \nu \qquad (10.33)$$

The optimized criterion, suggested by J. Brocas [47], is as follows. For a fixed V_{O_2}, the control variables V, V_T, $\mathcal{V}$ are chosen so as to minimise the energy consumption. The mathematical model is based on a simplified scheme [47] representing the thorax-lung system by a cylinder of constant section πr^2 and of variable height. The variation corresponds to the working of the respiratory muscles which act effectively as a piston. These muscles, of length l, contract by Δl. Where l_0 is the resting length, we have $l \in [0.6\ l_0,\ l_0]$. l_0 corresponds to the relaxation volume V_0, and $l = 0.6\ l_0$ corresponds to the maximum pulmonary capacity CT. We shall only consider ventilatory mechanics with passive expiration. A generalization to active expiration is possible and is studied in [47]. The inspiratory and expiratory muscles are represented schematically by a single muscle whose length is within the interval $[0.66\ l_0,\ 1.10\ l_0]$. The initial volume V can vary between V_R and CT. The model equations represent the dynamic and mechanical constraints of the ventilatory system.

a) Structure constraints

The pressure P_c needed to maintain the volume V is identified from experimental data.

$$P_c(u) = Iu^3 + Ju_2 + Ku + L \qquad (10.34)$$

where I, J, K, L are constants identified from the data.

The air passages have a resistance given by:

$$R(u) = (au + b)/(u + b) \qquad (10.35)$$

where a, b, c are identified from experimental data. In our model we use:

$$u = V + V_T/m, \quad m = 100 \qquad (10.36)$$

b) Dynamic constraints

$\dot{V}$ is due to the pressure difference $P_A - P_B$ between the alveolus (alveolar pressure P_A) and the atmosphere (pressure P_B). The following relation was chosen:

$$P_A - P_B = K_1 \dot{V}_1 + K_2 \dot{V}_2 \qquad (10.37)$$

where P_B may be set to zero (reference pressure).
The ventilated gas $\dot{V}$ is given by:

$$\dot{V} = V_T \cdot \nu \quad (\nu = \text{ventilation frequency}) \quad (10.38)$$

We also have a relation between $\dot{V}_{O_2}$ and $\dot{V}_A$:

$$\dot{V}_A = V_A \cdot \nu = V - V_D \nu$$

$$\dot{V}_{O_2} = \dot{V}_A (F_{IO_2} - F_{AO_2}) \quad \text{where}$$

F_{IO_2} = mean proportion of oxygen in the inspired air

F_{AO_2} = mean proportion of oxygen in the alveolar air

We thus obtain:

$$\dot{V} - V_D \nu = \dot{V}_{O_2} / (F_{IO_2} - F_{AO_2})$$

and
$$\nu = 1/V_D (\dot{V} - \dot{V}_{O_2} / (F_{IO_2} - F_{AO_2})) \quad (10.39)$$

c) Muscular constraints

The respiratory muscles exert a force F on the wall of the
thorax (see P. Nelson [53]). The following formula relates
the length, force and speed. It was identified from
experimental data in the literature:

$$F/F_0 = 1.67(l/l_0 - 0.4)/(1 + \tau v/l_0)$$

$$- 1/4 \ (\tau v/l_0)/(1 + \tau v/l_0) \quad (10.40)$$

where F is the force developed by the muscle at initial
length l and speed of contraction v ($v = \Delta l/\Delta t$) [53].
F_0 is the maximal force developed by the muscle at
speed $v = 0$ and initial length l_0 (isometric contraction).
$\tau = 0.1$ sec. is a time constant relevant to striated muscle.
In our particular problem, volumes correspond to length
in formula (10.40), and flow rates to speed. Equation
(10.40) becomes:

$$F/F_0 = 1/(1 + (0.1V_T)/t(H-V_0))$$

$$\cdot \ [1.67((H-V)/(H-V_0) - 0.4) + 1/4 \ 0.1V_T/t(H-V_0)]$$
$$(10.41)$$

with $H = \pi r^2 h$ = total volume of our cylinder.

Another useful parameter is t = duration of inspiration. It is related to the period T of the ventilatory cycle (T = 1/v) by:

$$t = 0.3T + 0.4$$

d)   The optimality criterion

We have to represent, mathematically, the power developed by the respiratory muscles to ensure a given $\dot{V}_{O_2}$. The energy required to transfer the surrounding air to the alveolus is:

$$\dot{V}_{O_2} (P_A - P_B) \qquad (10.42)$$

The respiratory muscles have to act in the following way to ensure the gaseous flow $\dot{V}$ corresponding to a fixed $\dot{V}_{O_2}$:

(i)   give power to overcome the elastic resistance:

$$\dot{V}P_C \qquad (10.43)$$

(ii) ensure a pressure difference between the atmosphere and the alveolus which requires the energy:

$$\dot{V}(P_A - P_B) \qquad (10.44)$$

The total mechanical energy given by the respiratory muscle is:

$$\dot{V}((P_A - P_B) + P_C) \qquad (10.45)$$

To provide this energy, the muscle needs a total energy of:

$$(1/\rho)\dot{V}((P_A - P_B) + P_C) \qquad (10.46)$$

where ρ represents the output of the system.

Now set: $\qquad\qquad \rho = \rho_{max} (F/F_0) \qquad\qquad (10.47)$

ρ_{max} is a parameter characterising the system.

When $F = F_0$, the muscle works at maximum efficiency (see (10.40)) and $\rho = \rho_{max}$. Then the minimization of the function:

$$y = (F/F_0)(1/\rho_{max}) \cdot \dot{V}((P_A - P_B) + P_C)/\dot{V}_{O_2}(P_A - P_B)$$
$$(10.48)$$

corresponds to our optimization problem.

The physiological system can be summed up as follows [30].

1. $P_C = IV_1^3 + JV_1^2 + KV_1 + L$

    with $I = 6.81277.10^9$, $J = 9.53447.10^6$, $K = 335185$,

    $L = -953.543$ and where $V_1 = V + V_T/m$

2. $R = (aV_1 + b)/(V_1 + c)$ with $a = -30798.9$, $b = 466.002$,

    $c = -1.91314.10^{-4}$, with V_1 as above.

3. $V_D = 0.1 V_T + 10^{-4}$

4. $P_A - P_B = R\dot{V} + K_2\dot{V}_2$ with $K_2 = 2.10^7$

5. $\nu = 1/V_D(\dot{V} - \dot{V}_{O_2}/(F_{IO_2} - F_{AO_2}))$; $F_{IO_2} - F_{AO_2} = 0.05$

6. $\dot{V} = \nu V_T$

7. $F/F_0 = 1/(1 + (0.1V_T)/t(H-V_0))$

    $\cdot [1.67((H-V)/(H-V_0) - 0.4) + 1/4 \; 0.1V_T/t(H-V_0)]$

    with $t = 0.3T + 0.4$ (10.49)

Furthermore the function $y(V,\dot{V})$ given by:

$$y(V,\dot{V}) = (F/F_0)(1/\rho_{max}) \cdot \dot{V}((P_A - P_B) + P_C)/\dot{V}_{O_2}(P_A - P_B)$$
$$(10.50)$$

has to be minimised as a function of V and $\dot{V}$.

The parameters P_A, P_C, R, $\dot{V}$, V, V_T, ν, V_D, F are unknown.

The previous numerical values were obtained using the M.K.S. system and the following data [30]:

$H = 14.14.10^{-3} \; m^3$ $F_{IO_2} - F_{AO_2} = 0.05$ $F_0 = 100 \; N$

$P_B = 0$ Pascal $K_2 = 2.10^7 \; kg \; m^{-4} \; s^{-1}$

$V_0 = 2.4.10^{-3} \; m^3$ $\rho_{max} = 0.2$

Parameters are defined as follows:

V = pulmonary volume (m^3)

$\dot{V}$ = gaseous flow $(m^3 \, s^{-1})$

V_T = tidal volume (m^3)

ν = ventilatory frequency (cycles s^{-1})

P_C = compliance pressure (Pascal)

P_A = alveolar pressure (Pascal)

R = air resistance (Pascal $m^3 \, s^{-1}$)

F = force developed by the muscle (Newton)

We also use numerical bounds for some unknowns:

$V \in [2.4.10^{-3}, \, 6.10^{-3}]$, $V \in [0.1.10^{-3}, \, 2.10^{-3}]$,

$\dot{V}_{O_2} \in [4.10^{-6}, \, 4.3.10^{-6}]$

Making the following change of variables:

$X = 10^3 V$ $\dot{X} = 10^3 \dot{V}$ $W = 2.10^4 V_{O_2}$

$B = 9\dot{X} - 10W$ $A = mBX + \dot{X}$

$\delta = 6.81277$ $\epsilon = 9.53447$ $\eta = 335.185$

$\mu = -953.543$ (10.51)

allows the functional y in (10.50) to be expressed as function of X and $\dot{X}$. We obtain:

$$y = J(X, \dot{X}) = (((466mB - 30.8A + K_2' \dot{X}(A - 0.2mB))B^3 m^3 \dot{X}$$

$$+ (A - 0.2mB)(\delta A^3 + \epsilon A^2 Bm + \eta AB^2 m^2 + \mu B^3 m^3))$$

$$. ((422.64 + D)\dot{X} - 469.6W + 35.22)))$$

$$/ (((466mB - 30.8A + K_2' \dot{X}(A - 0.2mB))B^3 m^3)$$

$$. ((567.7783 - 60.12X - D/4)\dot{X}$$

$$+ (66.8W - 5.01)X - 630.859W + 47.3144))$$

$$\text{(10.52)}$$

with $D = 1$, $K_2' = 20$, and $m = 100$.

(10.52) obviously depends only on X and $\dot{X}$.

to be added:

$$2.4 \leqslant X \leqslant 6, \quad 10W/9 \leqslant \dot{X} \leqslant 2 \qquad (10.53)$$

(10.52) and (10.53) pose an optimization problem with constraints which may be solved using a penalty technique (see Chapter 3). An optimization method (local or global) is needed. In [47], the Alienor method and the technique of Vignes [70] were employed. The numerical results [70] are very similar and show that, in this case, local methods converge towards the global minimum.

10.4 Conclusions

These two physiological examples show the power of the approaches using an optimization criterion. Many other biological problems can be studied using such techniques. Another example is the study of a complex system such as the gait [7]. A minimization criterion has to be introduced to ensure that the mathematical solution associated with the model is unique. The modelling of muscle also needs a supplementary criterion [11]. The cardiac system may be studied using energy minimization [51].

What is proved about the validity of these optimization criteria? We cannot ensure that they are valid from the biological point of view, but we can assert that they are compatible with the experimental data currently available about the phenomena being studied.

CHAPTER 11

ERRORS IN MODELLING

11.1 Compartmental modelling

Let us first consider the problem of the identification of a_i and λ_i from:

$$x(t) = \sum_{i=1}^{n} a_i \exp(\lambda_i t) \qquad (11.1)$$

where $x(t)$ is measured at $t = t_k$, $k = 1, \ldots, m$. Suppose, also, that the experimental error $\Delta x(t_k)$ on each measure $x(t_k)$ is known. More precisely, it is sufficient to have a bound for this error. It would be interesting to calculate the error on a_i and λ_i as a function of the error $\Delta x(t_k)$ of the experimental data. To obtain this relation, let us differentiate (11.1) with respect to a_i and λ_i. The following equation is obtained:

$$\Delta x(t) = \sum_{i=1}^{n} \exp(\lambda_i t)(\Delta a_i)) + \sum_{i=1}^{n} a_i t \exp(\lambda_i t)(\Delta \lambda_i)) \quad (11.2)$$

Substituting (11.2) for $t = t_k$ leads to the linear algebraic system:

$$\Delta x(t_k) = \sum_{i=1}^{n} \exp(\lambda_i t_k)(\Delta a_i)) + \sum_{i=1}^{n} a_i t_k \exp(\lambda_i t_k)(\Delta \lambda_i)) \qquad (11.3)$$

where $k = 1, \ldots, m$, which includes 2n unknowns (the Δa_i and the $\Delta \lambda_i$) and m equations.

In order to obtain one solution, at most, we assume that $m \geq 2n$. Then the linear system (11.3) can be solved using a numerical method for the solution of linear algebraic systems [20] (Gauss method, ...) when $m = 2n$ and introducing a functional to minimise it if $m > 2n$. This functional J, may be:

$$J = \sum_{k=1}^{m} [\Delta x(t_k) - \sum_{i} \exp(\lambda_i t_k)(\Delta a_i))$$

$$- \sum_{i} a_i t_k \exp(\lambda_i t_k)(\Delta \lambda_i)]^2 \qquad (11.4)$$

Its optimum is found by writing:

$$\partial J / \partial \Delta a_i = 0 = \partial J / \partial \Delta \lambda_i \qquad (11.5)$$

This system has 2n equations and 2n unknowns.

In spite of its simplicity this process is sometimes difficult because of numerical singularity of the matrix corresponding to the system (11.3).

When studying simple models (2 or 3 compartments) we know that exchange parameters are explicitly given as a function of a_i and λ_i. An adaptation of the previous technique may be used for estimating the errors in the k_{ij} parameters.

Another practical method used by A. Guillez (and probably many other people!) when treating pharmacological problems is to perturb the experimental data (within the admissible range associated with the experimental errors). The solution (a_i, λ_i for example) is calculated for each set of perturbed data and it is easy to use statistical methods to determine the error on each parameter. In fact, the mean and variance are readily obtained from the numerical results. In practice, many pharmacokinetic problems (from the Sandoz laboratories) were solved, and the numerical error evaluated using this empirical approach. The actual results were satisfactory [13].

A more sophisticated technique (called permutation perturbation method) was given by J. Vignes in [71]. In some ways it is a generalization of the previous method, taking the errors of calculation into account. Firstly, J. Vignes makes the distinction between two kinds of error [71]:

a) methods giving an exact result are called direct (numerical) methods. This is the case in the technique of linear algebra.

b) methods which which give approximate results are called approximate (numerical) methods. Results then have a method error.

When these methods (direct or approximate) are used on a digital computer the representation space is necessarily discrete and consequently the arithmetic precision is limited. Hence, when a digital computer is used there are two outcomes:

(i) a direct method gives a result with only a computational error, while

(ii) an approximate method has both a method error and a computing error, in the result.

Now let us give a precise evaluation for direct numerical methods. An exact numerical method is a sequence of operations (addition, subtraction, multiplication and division). Some numerical result, R, corresponds to the exact result, r. However, the operations may be permuted in their sequence in time (in some computations, all of them) because of the algebraic properties of associativity and commutativity. Therefore, if C_{op} is the total number of combinations corresponding to all the different permutations of the operations, there are C_{op} data-processing procedures that can give a value for R.

Let us consider any data-process that is an image of the algebraic procedure. Since each operator produces an assignment error, the result has an error. This may be a deficiency (when the computer uses truncated arithmetic) or, equally legitimately, an excess. So, any data-processing operation has two results - one deficient and one in excess.

Suppose that the data-processing procedure has k elementary operations. Then this procedure yields not a single result for R, but 2^k, all equally representative of r [71].

Now we apply the perturbation method to each data-processing algorithm provided by the permutation method. We obtain a set $\mathcal{R}$ of results whose cardinal is:

$$\text{Card } \mathcal{R} = 2^k C_{op}$$

A simple statistical treatment easily gives C, the number of significant decimal digits. Taking $R_0 \in \mathcal{R}$ corresponding to the algebraic algorithm, translated without using permutation or perturbation, and calling the mean and standard deviation of the elements of $\mathcal{R}$, $\bar{R}$ and δ, respectively, we obtain C by the equation:

$$\epsilon / |R_0| = 10^{-C} \text{ with } \epsilon = ((R_0 - \bar{R})^2 + \delta^2)^{1/2} \qquad (11.6)$$

This method giving the number of significant decimal digits is excellent from the theoretical point of view. Unfortunately, there are many extra calculations and so its use may be prohibited on micro-computers. In practice, the analysis of errors is made using sensitivity functions [29].

11.2 Sensitivity analysis

Let us consider a second order differential equation:

$$F(\ddot{x}, \dot{x}, x, t, p) = 0 \qquad (11.7)$$

where p is a parameter to be identified from the experimental data. The solution of (11.7) depends on t and p: x = x(t,p)

A perturbation Δp on p transforms the differential equation (11.7) which becomes:

$$F(\ddot{x}, \dot{x}, x, p + \Delta p) = 0 \qquad (11.8)$$

The sensitivity parameter u(t,p) is then defined by:

$$u(t,p) = \lim_{\Delta p \to 0} (x(t, p + \Delta p) - x(t,p))/\Delta P = \partial x/\partial p(t,p) \qquad (11.9)$$

The coefficient u measures the effect of a perturbation ΔP on the solution x(t). Differentiating the differential equation (11.7) with respect to p leads to:

$$\partial F/\partial \ddot{x} \cdot \partial \ddot{x}/\partial p + \partial F/\partial \dot{x} \cdot \partial \dot{x}/\partial p + \partial F/\partial x \cdot \partial x/\partial p + \partial F/\partial p = 0 \qquad (11.10)$$

Setting:

$$\dot{u} = \partial/\partial t \cdot \partial x/\partial p = \partial \dot{x}/\partial p \text{ and } \ddot{u} = \partial^2/\partial t^2 \cdot \partial x/\partial p = \partial \ddot{x}/\partial p$$

gives:

$$\partial F/\partial \ddot{x} \cdot \ddot{u} + \partial F/\partial \dot{x} \cdot \dot{u} + \partial F/\partial x \cdot u = \partial F/\partial p \qquad (11.11)$$

The solution of (11.11) uses the sensitivity function u(t,p).

Now let us consider a compartmental example, such as that shown in Figure 15, where:

G = gastro-intestinal tract, B = blood and U = urine.

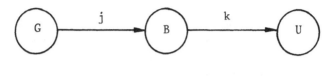

FIG. 15

which is described by the differential system:

$$\dot{G} = - jG \qquad G(0) = a$$

$$\dot{B} = jG - kB \qquad\qquad\qquad (11.12)$$

$$\dot{U} = kB \qquad B(0) = U(0) = 0$$

Expressing the effect of a perturbation of j on the state variables G, B, U by differentiating (11.12) with respect to j gives:

$$\partial^2 G/\partial t \partial j = - j(\partial G/\partial j) - G$$

$$\partial^2 B/\partial t \partial j = j(\partial G/\partial j) + G - k(\partial B/\partial j) \qquad (11.13)$$

$$\partial^2 U/\partial t \partial j = k(\partial B/\partial j)$$

If we set:

$$\partial G/\partial j = x, \quad \partial B/\partial j = y \quad \text{and} \quad \partial U/\partial j = z \qquad (11.14)$$

we obtain the following equations (sensitivity equations):

$$\dot{x} = -jx - G$$

$$\dot{y} = jx + G - ky \qquad\qquad (11.15)$$

$$\dot{z} = ky$$

The simultaneous solution of systems (11.12) and (11.15) gives the sensitivity functions.

They can be used as weighting coefficients in the functionals which are used for identifying unknown parameters. For example, if we have to minimise:

$$\sum_{j=0}^{n} [G(t_j) - G^c(t_j)]^2$$

where $G(t_j)$ is the experimental data and $G^c(t_j)$ is the value calculated from (11.12), it is possible to introduce the new functional:

$$\sum_{k=0}^{n} [G(t_k) - G^c(t_k)]^2 x(t_k) \qquad (11.16)$$

taking into account the sensitivity of G with respect to the parameter t_k.

More generally, let us introduce the general non-linear differential system which corresponds to a biological model:

$$\dot{X} = F(X(t),P)$$
$$\qquad\qquad\qquad\qquad (11.17)$$
$$X(0) = X_0$$

where F is known or identified; X(t) is the state vector and P is a paramter vector (corresponding, for example, to the k_{ij} in a compartmental system).

The sensitivity functions are defined by:

$$s_i(t) = \partial X(t)/\partial p_i \qquad (11.18)$$

where p_i is the ith component of the vector p.

Since the system (11.17) is non-linear, it is not possible to find the sensitivity function $s_i(t)$ directly. We can use an integral formulation for (11.17):

$$X(t) = X(0) + \int_0^t F(X(s),p)ds \qquad (11.19)$$

Differentiating (11.19) gives:

$$\partial X/\partial p_i (t) = \partial X/\partial p_i(0) + \int_0^t (\partial F/\partial X \cdot \partial X/\partial p_i + \partial F/\partial p_i)ds$$
$$\qquad\qquad\qquad\qquad (11.20)$$

Differentiating again, with respect to time t, gives:

$$d/dt\ s_i(t) = \partial F/\partial X\ s_i(t) + \partial F/\partial p_i \qquad (11.21)$$

Indeed, $X(0)$ fixed gives $s_i(0) = 0$.

Then the sensitivity functions are solutions of the differential system (11.21) which is well-defined because the function F is known. It may be solved using, for example, the Runge-Kutta technique [24]. In [58], sensitivity functions associated with the hormonal system described in Chapter 5 can be found.

These functions allow us to find some error bounds on the parameters to be identified. As before, we notice that a perturbation Δp on p involves a perturbation on $X(t)$ denoted by ΔX and given by:

$$\Delta X(t) = \sum_i \partial X / \partial p_i (t) \Delta p_i = \sum_i s_i(t) \Delta p_i \qquad (11.22)$$

Since $\Delta X(t)$ is known (experimental error), the solution of $\Delta p = (\Delta p_i)$ may be found by solving the linear system:

$$\Delta X = S \Delta p \qquad (11.23)$$

where S is the sensitivity matrix.

Of course, (11.23) may be discretised by considering $t = t_i$, $i = 1, \ldots, m$ involving an algebraic linear system giving the $\Delta X_j(t_i)$ with $X = (X_j)$. This method works when $S.S^T$ is a well-conditioned matrix.

Another approach for evaluating the errors may be developed. Let us denote the error between X experimental (X_e) and X calculated (X_c) by δX. Then, we have:

$$\delta X = \int_0^T (X_e - X_c)^2 dt \qquad (11.24)$$

where X_e and X_c are vectors. Then if δP is the error on P, we have the relation:

$$\| \delta P \| \leqslant ((\delta X) \text{ cond } (S.S^T) / \| S.S^T \|)^{(1/2)} \qquad (11.25)$$

where, by definition [20], the conditioning of a matrix C is expressed by:

$$\text{cond } (C) = \| C \| \cdot \| C^{-1} \| \qquad (11.26)$$

In (11.25) $\| \delta P \|$ represents the usual vector norm and $\| S.S^T \|$ is the associated matrix norm. Remember that [20] for a linear algebraic system:

$$CX = B \qquad (11.27)$$

with matrix C and vector B given, and X unknown. An error
bound is given by:

$$\|\delta A\| \leq \|A\| \, \text{cond (C)}. \, (\|\delta B\|)/\|B\| \qquad (11.28)$$

This implies that the solution found is meaningless when
there has been extensive conditioning.

Remark It would be possible to use the sensitivity function
to determine the best measurement times, t_i, to ensure
optimal values of $s_i(t)$. The mathematical problem is to
solve a minimization system. The reader may easily do this
as an exercise.

Note Errors in modelling are treated in [29].

CHAPTER 12

OPEN PROBLEMS IN BIOMATHEMATICS

12.1 Biological systems with internal delay

Some biological phenomena cannot be explained by classical compartmental anlysis. This is the case when the data is a concentration curve having two maxima. A usual compartmental model is not satisfactory and a new approach is necessary. The literature [28], [73] is still very poor in methods which can be adapted for complex phenomena.

We shall present some ideas which are useful in the study and solution of such problems. However, there is still a great deal of work, both practical and theoretical, to be done. The identification of control problems is much more difficult than in the linear case where there is no delay. The following mathematical developments are associated with a drug called Estulic, which has been studied by the Sandoz laboratories. The following data was kindly given by Mrs. Guerret and Mrs. Lavene of Sandoz. The numerical study was done by A. Guillez [37].

The experimental data is concentration in the blood, $C_1(t)$ measured at $t = t_j$, $j = 1, \ldots, m$. A reasonable model is shown in Figure 16.

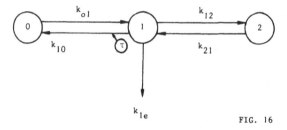

FIG. 16

where τ corresponds to a delay.

The first mathematical problem is to identify the k_{ij} and from the experimental data. We shall study the system represented by Figure 16 in intervals of length τ.

(i) The retarded function does not exist on $[0, \tau]$, and we

217

obtain the differential system:

$$\dot{C}_{01}(t_1) = - k_{01}C_{01}(t_1)$$

$$\dot{C}_{11}(t_1) = k_{01}C_{01} - (k_{12} + k_{1e})C_{11} + k_{21}C_{21}$$

$$\dot{C}_{21}(t_1) = k_{12}C_{11} - k_{21}C_{21}$$

(12.1)

$$C_{01}(0) = C_{00}, \quad C_{11}(0) = C_{21}(0) = 0, \quad t_1 \in [0,\tau]$$

where the second index 1 corresponds to the first interval, and D = given oral dose.

We know how to identify this system. Firstly, we identify the a_{1i} and λ_i (i = 1,2,3) of the formula:

$$C_{11}(t_1) = a_{11} \exp(-\lambda_1 t_1) + a_{12} \exp(-\lambda_2 t_1) + a_{13} \exp(-\lambda_3 t_1)$$

(12.2)

and we can set $\lambda_1 < \lambda_2 < \lambda_3$. The relation $C_{11}(0) = 0$ implies $a_{13} = - a_{11} - a_{12}$. Previously we chose the largest value of λ_i for k_{01}, that is to say, $k_{01} = \lambda_3$. Writing the relations between the unknowns leads to:

$$k_{21} = ((a_{11}\lambda_2 + a_{12}\lambda_1)k_{01} + a_{13}\lambda_1\lambda_2)$$

$$/ ((k_{01} - \lambda_1)a_{11} + (k_{01} - \lambda_2)a_{12}))$$

(12.3)

$$k_{1e} = \lambda_1\lambda_2/k_{21}, \quad k_{12} = \lambda_1 + \lambda_2 - k_{21} - k_{1e};$$

$$C_{00} = C'_{11}(0)/k_{01}$$

(12.4)

$$C_{01}(t_1) = C_{00} \exp(-k_{01}t_1)$$

(12.5)

$$C_{21}(t_1) = k_{12}[(a_{11}\exp(-\lambda_1 t_1))/(k_{21} - \lambda_1))]$$

$$+ (a_{12}\exp(-\lambda_2 t_1))/(k_{21} - \lambda_2))$$

$$+ (a_{13}\exp(-k_{01}t_1))/(k_{21} - k_{01}))$$

(12.6)

Now we consider the second interval where the retarded function is $C_{11}(t_1 - \tau)$. The following differential system is satisfied:

$$\dot{C}_{02}(t_1) = - k_{01}C_{02}(t_1) + k_{10}C_{11}(t - \tau)$$

$$\dot{C}_{12}(t_1) = k_{01}C_{02}(t_1) - (k_{12} + k_{1e})C_{12}(t_1)$$

$$+ k_{21}C_{22}(t_1) - k_{10}C_{11}(t_1 - \tau)$$

(12.7)

$$\dot{C}_{22}(t_1) = k_{12}C_{12} - k_{21}C_{22}$$

$$C_{12}(\tau) = C_{11}(\tau), \quad C_{22}(\tau) = C_{21}(\tau), \quad C_{02}(\tau) = C_{01}(\tau)$$

$$C'_{02}(\tau) = C'_{01}(\tau), \quad C'_{12}(\tau) = C'_{11}(\tau), \quad C'_{22}(\tau) = C_{21}(\tau)$$

The system (12.7) may be considered to be the system (12.1) excited by the vector $(k_{10}C_{11}(t - \tau), -k_{10}C_{11}(t_1 - \tau), 0)$. This vector has the same characteristic values as the differential system. Thus we may write:

$$C_{i2}(t_1) = \sum_{j=1}^{3} (b_{ij}t_1 + C_{ij})\exp(-\lambda_j t_1); \quad \lambda_j = \lambda_1, \lambda_2, \lambda_3, \cdot \cdot$$

$$i = 0, 1, 2$$

(12.8)

The expression (12.8) contains (for each i) six parameters b_{ij} and C_{ij}. Identification is usually impossible because of the small amount of experimental data on $(\tau, 2\tau)$.

More generally, the general system on $((n-1)\tau, n\tau)$ is represented by the following:

$$\dot{C}_{0n}(t_1) = - k_{01}C_{0n}(t_1) + k_{10}C_{1,n-1}(t_1 - \tau)$$

$$\dot{C}_{1n}(t_1) = k_{0n}C_{0n}(t_1) - (k_{12} + k_{1e})C_{1n}(t_1)$$
$$- k_{10}C_{1,n-1}(t_1 - \tau)$$

$$\dot{C}_{2n}(t_1) = k_{12}C_{1n}(t_1) - k_{21}C_{2n}(t_1) \quad \text{and with the conditions}$$

$$C_{in}((n-1)\tau) = C_{i,n-1}((n-1)\tau), \quad \dot{C}_{i,n-1}((n-1)\tau) = \dot{C}_{in}((n-1)\tau)$$

(12.9)

The exact solutions in $((n-1)\tau, n\tau)$ are given by:

$$C_{in}(t_1) = \sum_{j=1}^{3} P_{j,n-1}(t_1)\exp(-\lambda_j t_1), \quad i = 0, 1, 2 \qquad (12.10)$$

where $P_{j,n-1}$ is a polynomial function of degree $(n-1)$.

The expressions (12.10) contain $3n$ coefficients and thus cannot be identified in practice.

To overcome this, let us transform the system (12.9) into a recurrent equation. Using the operator $D = d/dt_1$ we obtain:

$$C_{0n}(t_1) = k_{10}C_{1,n-1}(t_1 - \tau)/(D + k_{01})$$

$$C_{2n}(t_1) = k_{12}C_{1n}(t_1)/(D + k_{21}) \qquad (12.11)$$

Substituting (12.11) in the last differential system (12.9) gives:

$$(D + k_{12} + k_{1e} - k_{21}k_{12}/(D + k_{21}))C_{1n}$$

$$= - [k_{10}D/(D + k_{01})]C_{1,n-1}(t - \tau) \qquad (12.12)$$

The recurrent equation relating C_{1n}, $C_{1,n-1}(t_1 - \tau)$ is thus the following:

$$(D - k_{01})(D - \lambda_1)(D - \lambda_2)C_{1n}(t_1)$$

$$= - k_{10}(D + k_{21})DC_{1,n-1}(t_1 - \tau) \qquad (12.13)$$

with:
$$C_{1n}((n-1)\tau) = C_{1,n-1}((n-1)\tau)$$

$$\overset{\bullet}{C}_{1,n-1}((n-1)\tau) = \overset{\bullet}{C}_{1n}((n-1)\tau)$$

A numerical identification method results from these.

A first approximation to τ (denoted by τ_1) comes from formula (12.2). In fact, τ may be considered as the abcissa of the first maximum calculated from it. An improvement is obtained by considering the second interval $[\tau_1, 2\tau_1]$. The following algebraic system is formed by choosing two points A_1 and B_1 in this interval at which the function C_1 is measured:

$$C_{1A_1} = \alpha_1 \, \text{Log}^2(1 + t_{1A_1}) + \beta_1 \, \text{Log}(1 + t_{1A_1}) + \gamma_1$$

$$C_{1B_1} = \alpha_1 \, \text{Log}^2(1 + t_{1B_1}) + \beta_1 \, \text{Log}(1 + t_{1B_1}) + \gamma_1$$

$$\alpha_1 \, \text{Log}^2(1 + \tau_1) + \beta_1 \, \text{Log}(1 + \tau_1) + \gamma_1$$

$$= a_{11} \, \exp(-\lambda_1 \tau_1) + a_{12} \, \exp(-\lambda_2 \tau_1) + a_{13} \, \exp(-\lambda_3 \tau_1)$$

$$2\alpha_1 \, \text{Log}(1 + \tau_1) + \beta_1 = - (1 + \tau_1)(\lambda_1 a_{11} \, \exp(-\lambda_1 \tau_1)$$

$$+ \lambda_2 a_{12} \, \exp(-\lambda_2 \tau_1) + \lambda_3 a_{13} \, \exp(-\lambda_3 \tau_1))$$

$$(12.14)$$

The system (12.14) is obtained by taking the following approximate structure for C_{12}:

$$C_{12} = \alpha_1 \, \text{Log}^2(1 + t_1) + \beta_1 \, \text{Log}(1 + t_1) + \gamma_1$$
$$C'_{12} = (2\alpha_1 \, \text{Log}(1 + t_1) + \beta_1)/(1 + t_1)$$
$$(12.15)$$

(12.14) is simply (12.15) with $t_1 = \tau_1$, then with $t_1 = t_{1A_1}$, t_{1A_2}. The numerical solution of (12.14), containing four unknowns α_1, β_1, γ_1, τ_1 gives a new value of τ, denoted by τ_2.

The process can be iterated and an improved value of τ_2 may be obtained by considering the interval $[2\tau_2, 3\tau_2]$. The process is continued on each interval $[(n-1)\tau_{n-1}, n\tau_{n-1}]$, $n = 1, \ldots, N$. With the last estimation τ_N we choose $\tau = \tau_N$ and all the coefficients (α, β, γ) are recalculated. A good sequence (C_{1n}) is obtained. It is defined by (12.15) on each interval $[(n-1)\tau, n\tau]$. Then the recurrent relation (12.13) is used on each interval. It gives an algebraic system allowing determination of the definitive values of λ_1, λ_2, λ_3, k_{10}, k_{21}.

Remark Of course, approximations other than logarithmic polynomials may be chosen to approximate $C_1(t)$. Other mathematical approaches may be possible for explaining concentration curves with at least two extrema (open problem!). For instance, an integral equation with time lag could be set up:

$$C_1(t) = \int_0^T K(s,t) \, C_1(s - \tau) ds + f(t) \qquad (12.16)$$

But what biological meaning could be attached to K and f?   Recently, A. Guillez proposed using second order differential systems to describe two extrema.   These models are justified by biological considerations (to appear).

With the previously-described technique the following data from Estulic (Sandoz Laboratories) were used:

t(h)	$C_1(\mu g/l)$	t	C_1	t	C_1
0	0	3	3.23	15	1.72
0.5	2.94	4	2.74	19	1.13
1.0	4.86	6	2.90	24	0.80
1.5	4.69	8	3.65	37	0.45
2.0	4.13	11	2.60		

A first approximation is made using the values:

$$t = 0.5, \ 1, \ 1.5$$

$$C_1 = -14.5566 \ \text{Log}^2(1 + t) + 22.6661 \ \text{Log}(1 + t) - 3.8572$$

$$(12.17)$$

The relation $C_1(t) = 0$ gives:

$$t_0 = 0.2147 \ \text{hours} = 12 \ \text{min} \ 53 \ \text{sec}$$

and corresponds to the initial time lag.   Thus the appropriate time t_1 is:

$$t_1 = t - 0.2147$$

The exponential approximation is found by using the global optimization technique (Alienor).   We obtain:

$$C_{11}(t_1) = 2.8014 \ \text{exp}(-0.0377t_1) + 7.3763 \ \text{exp}(-0.853t_1)$$

$$- 10.1777 \ \text{exp}(-2.352t_1)$$

$$(12.18)$$

which is valid up to 4 hours, and thus $T_1 = 4$ hours was chosen.

From t = 8 hours to t = 37 hours we find:

$$C_{13}(t_1) = 9.3225 \exp(-0.21t_1) + 2.6954 \exp(-0.05t_1)$$
(12.19)

and the correspondence of this formula to the numerical data is shown in the following table:

t	8	11	15	19	24	37
C_{13}	3.65	2.60	1.72	1.13	0.80	0.45
C_{13} calc	3.67	2.55	1.71	1.24	0.89	0.43

At the time values shown the comparison between C_{11} calculated and the actual graph is as follows:

t_1	3.733	3.983	4.233	4.483	4.733	4.983	5.233	5.733
C_{11} calc	2.74	2.66	2.59	2.53	2.47	2.43	2.38	2.31
C_1 graph	2.74	2.67	2.60	2.53	2.50	2.50	2.68	2.90
t	4.00	4.25	4.50	4.75	5.00	5.25	5.50	6.00

Recall that t = 4 hours corresponds to t_1 = 3.733.

The break in the curve is in the interval [4.483,4.733]. We may take T_1 = 4.5 as a first approximation. The function C_{12} which satisfies the boundary conditions is equal to:

$$C_{12}(t_1) = 0.019584\ t_1^4 - 0.57192\ t_1^3 + 6.01188\ t_1^2$$

$$- 26.7253\ t_1 + 45.1315$$
(12.20)

In this case the polynomial expansion is chosen because the coefficients of the logarithmic one are too large. From the expression $C_{11}(t_1)$, the following parameters and functions are derived:

$k_{01} = 2.352$, $k_{21} = 0.3394$, $k_{1e} = 0.09476$, $k_{12} = 0.4566$

$C_{00} = 7.4625$, $V = 268.0$ l (volume of compartment 1)

$C_{01} = 7.4625 \exp(-2.352t_1)$

$C_{21}(t_1) = 4.239 \exp(-0.0377t_1) - 6.558 \exp(-0.853t_1)$

$$+ 2.319 \exp(-2.352t_1) \qquad (12.21)$$

We can now use the recurrent relation between $C_{12}(t_1)$ and $C_{11}(t - \tau)$ and between $C_{13}(t_1)$ and $C_{12}(t_1 - \tau)$. We find:

$$\tau = 4.505, \quad k_{10} = 11.305 \ h^{-1}$$

The variation on τ is too small to modify the other coefficients. Then the process is continued and leads to the final result:

$k_{01} = 2.352$, $k_{12} = 0.4566$, $k_{1e} = 0.09476$, $k_{21} = 0.3394$

$k_{10} = 11.305$, $\tau = 4.505h$, $V = 268$ l. $t_0 = 0.267h$
$$(12.22)$$

Let us recall and complete the formulae:

On $[0, \tau]$ we have:

$C_{11}(t_1) = 2.8014 \exp(-0.0377t_1) + 7.3763 \exp(-0.853t_1)$
$$- 10.177 \exp(-2.352t_1)$$

$C_{01}(t_1) = 7.4625 \exp(-2.352t_1)$

$C_{21}(t_1) = 4.239 \exp(-0.0377t_1) - 6.558 \exp(-0.853t_1)$
$$+ 2.319 \exp(-2.352t_1)$$

On $[\tau, 2\tau]$ we have:

$C_{12}(t_1) = 0.019583 \ t_1^4 - 0.57191 \ t_1^3 + 6.01188 \ t_1^2$
$$- 26.72533 \ t_1 + 45.1315$$

$C_{02}(t_1) = 16.2148 \exp(-0.0377t_1) + 2595.3686 \exp(-0.853t_1)$
$$- (3190031.618t_1 - 11060140.56) \exp(-2.352t_1)$$

$C_{22}(t_1) = 0.026345 \ t_1^4 - 1.07988 \ t_1^3 + 17.63303 \ t_1^2$
$$- 139.861 \ t_1 + 472.8 - 504.1118 \ \exp(-2.352t_1)$$

and we obtain, at last, on $[2\tau, 3\tau]$:

$$C_{13} = 9.3225 \exp(-0.21t_2) + 2.6954 \exp(-0.05t_2)$$

$$C_{03}(t_2) = 0.219352 \, t_2^4 + 2.851 \, t_2^3 - 14.356 \, t_2^2$$

$$+ 33.5613 \, t_2 - 2.152 + 12.73 \exp(-2.352t_2)$$

$$C_{23}(t_2) = 32.89529 \exp(-0.21t_2) + 4.25266 \exp(-0.05t_2)$$

$$- (74.27093t_2 \quad \exp(-0.3394t_2)$$

If $t_1 \geqslant 3\tau$, we have:

$$C_{0n}(t_1) = 11.2093 \exp(-0.11t_1) + 1.4667 \exp(-0.05t_1)$$

$$+ 1.7048 \cdot 10^{15} \exp(-2.352t_1)$$

12.2 Biological systems involving retroaction

Retroaction arises in many biological systems, and although there have been many attempts to study such systems [39] there is no general theory for their modelling. We will give an example developed by A. Guillez and S. Malet (Medimat Laboratory) in unpublished work. It refers to oocytes, the giant cells which become eggs after fertilisation. Amphibian oocytes from Xenopus [54] will be considered. In the ovary the oocytes are in a physiologically stable state at the end of their growth. They may remain in this state for a long time, and then develop further when laid. Oocytes I are transformed to oocytes II when they are laid. This is called the meiotic maturation and is due to progesterone (PG). From a biochemical point of view there are two phases:

- one depending on the cyclic AMP (cAMP)

- another independent of cAMP.

cAMP is a hormone playing a role in many biological mechanisms. It is made by another enzyme, adenilate cyclase (AC), and degraded by yet another, phosphodiesterase (PDE), as shown in the diagram:

The progesterone could act by decreasing the cAMP concentration [54]. Two key steps in the meiotic maturation of the Xenopus oocyte involve protein phosphorylation-dephosporylation [54].

- A decrease, in ovo, of the level of the free catalytic (C) sub-unit of the cAMP-dependent protein kinase (R_2C_2) initiates meiotic maturation.

- A burst of cAMP-independent phosphorylation occurs at the time of breakdown of the nuclear envelope, whether the maturation is triggered by progesterone or by micro-injection of MPF (maturation promoting factor). The following balance relation holds:

$$R_2C_2 + 4 \text{ cAMP} \rightleftharpoons 2R(\text{cAMP})_2 + 2C \qquad (12.23)$$

in which the protein kinase has two sub-units: R (regulator) and C (catalyst).

From the previous considerations a mathematical model for oocyte maturation may be proposed. First we represent the phenomena by a simplified diagram:

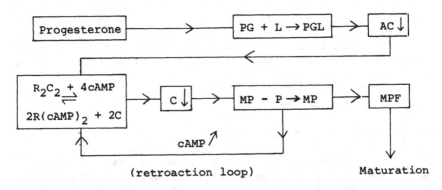

where L = liquid receptor; MP - P $\rightarrow$ MP = dephosphorylation, and MPF = maturation promoting factor.

We assume that the progesterone concentration $x_1(t)$ decreases according to the classical law $x_1(t) = x_0 \exp(-\alpha t)$ where x_0 is the concentration at time $t = 0$. The combination PGL catalysing the formation of PG comes from the reaction $PG + L \rightarrow PGL$. We assume it is mononuclear and let l_0 be the initial quantity of L. Then at time t there is $x_1(t) = [PG]$ and the concentration of L in combination is equal to:

$$\int_0^t x_1(u)\,du = (x_0/\alpha)[1 - \exp(-\alpha t)]$$

The remaining concentration is $l_0 - (x_0/\alpha)(1 - \exp(-\alpha t))$. But the rate parameter k depends on C:

$$k = Y(C_0) \cdot k_1(C) \qquad (12.24)$$

where Y is the Heaviside function.

For x, the PGL concentration, we obtain:

$$\dot{x} = Y(C_0) \cdot k(C) x_1 (l_0 - (x_0/\alpha)(1 - \exp(-\alpha t))) \qquad (12.25)$$

Let a_0 be the initial concentration of adenylate cyclase. The reaction involves a decrease of (AC) whose magnitude is completely unknown. Now consider the balance relation:

$$R_2C_2 + 4 \text{ cAMP} \underset{k_2}{\overset{k_1}{\rightleftharpoons}} 2R(cAMP)_2 + 2C$$

Note the balance concentrations:

$[C] = C_e$; $[RcAMP] = r_e$; $[R_2C_2] = a_e$ and $[cAMP] = b_e$.

Then the actual concentrations are:

$[C] = C$; $[RcAMP] = r_e + C - C_e$; $[R_2C_2] = a_e - (C - C_e)/2$;

$[cAMP] = b_e - 2(C + C_e)$

The differential equation representing the variation of C is:

$$\dot{C} = k_1(r_e + C - C_e)^2 c^2 - k_2(a_e - (C - C_e)/2)(b_e - 2(C + C_e))^4 \qquad (12.26)$$

From biological results the quotient $K = k_2/k_1$ is known and constant. Also, we know intervals within which k_1 and k_2 fall. Summing up, we have the following differential system:

$$\dot{x} = Y(C_0).k(C)x_1(l_0 - (x_0/\alpha)(1 - \exp(-\alpha t)))$$

$$\dot{C} = k_1((r_e + c - C_e)^2 c^2 \qquad (12.27)$$

$$- K(a_e - (C - C_e)/2)(b_e - 2(C + C_e))^4)$$

If we know the experimental curves for x and C, that is $x = f(t)$, $C = g(t)$, the unknown parameters of system (12.27) can be identified by solving the optimization problem:

$$\text{Min } ((x_c - x_e)^2 + (C_c - C_e)^2) \qquad (12.28)$$

where x_c and C_c are calculated functions obtained by solving (12.27) by the Runge-Kutta technique [24] with fixed parameters. x_e and C_e are the experimental functions.

In conclusion, we have obtained a simple mathematical model explaining oocyte meiotic maturation. Some hypotheses have to be formed in order to obtain a mathematical model which can then be studied using classical numerical methods (identification, simulation). Because of the lack of experimental data it was not possible to take into account the full complexity of the phenomenon. We obtained an approach that is global and may be useful.

More generally, we can say that many biological systems [53] are controlled by retroaction. It is, in fact, true for all systems where the nervous system plays a role.

12.3 Action of two (or more) drugs in the human organism

Combinations of drugs are frequent in human therapy. There is no general mathematical theory for this problem. The following ideas could be developed for its solution.

Suppose that two drugs are used together and that pharmacological studies give data for each drug separately, and for their interaction. In general the experimental data comes from the measurement of concentrations in the blood.

A compartmental model may be proposed for each dose:

$$\dot{x}_i = A_i x_i \qquad i = 1,2 \qquad (12.29)$$

where x_1 is the vector corresponding to the first drug and x_2 to the second one.

The first possibliity is that the drugs act independently, even when administered together. In this case the mathematical study is simple and can be reduced to the study of two independent compartmental systems such as (12.29). The most likely possibility involves interaction between the drugs and their effects. The system (12.29) has to be modified to take agonisms and antagonisms into account.

One of many possible approaches is to suppose that there is a relation between the plasma concentrations. To be more precise, let us consider two simple models associated with each drug:

$$\dot{x}_1 = -(k_{12} + k_e)x_1 + k_{21}x_2$$
$$\dot{x}_2 = k_{12}x_1 - k_{21}x_2$$

(12.30)

for the first drug, and:

$$\dot{y}_1 = -(K_{12} + K_e)y_1 + K_{21}y_2$$
$$\dot{y}_2 = K_{12}y_1 - K_{21}y_2$$

(12.31)

for the second.

An interaction can be modelled by the supplementary relation:

$$f(x_1,y_1) = 0 \quad \text{or} \quad x_1 = F(y_1) \qquad (12.32)$$

where f or F has to be identified from experimental data corresponding to the association of the drugs. The exchange parameters k_{ij}, and K_{ij} are identified using data for each drug (separately). This pre-supposes that the k_{ij} and K_{ij} are independent. If this were not true, the identification of the model would become even more difficult.

There is no general method available for choosing between f and F. A general study of concrete problems in pharmacokinetics could define the mathematical structure of f and F. Of course, optimal control problems may be associated with (12.30), (12.31) and (12.32). They correspond to the definition of optimal therapeutics.

For example, if we want to have, as closely as possible:

$$x_1(t) = a \quad , \quad y_1(t) = b \tag{12.33}$$

the following criterion can be introduced:

$$J = \int_0^T (x_1(t) - a)^2 dt + \lambda \int_0^T (y_1(t) - b)^2 dt \tag{12.34}$$

where λ is a constant chosen according to the importance of each sub-criterion. The functional J has to be minimised in terms of $u(t)$, $v(t)$ where $u(t)$, $v(t)$ are injections into compartment 1 of the first and second drugs, respectively. The state equations associated with (12.34) are as follows:

$$\dot{x}_1 = - (k_{12} + k_e)x_1 + k_{21}x_2 + u(t)$$

$$\dot{x}_2 = \quad k_{12}x_1 - k_{21}x_2$$

$$\dot{y}_1 = - (K_{12} + K_e)y_1 + K_{21}y_2 + v(t) \tag{12.35}$$

$$\dot{y}_2 = \quad K_{12}y_1 - K_{21}y_2$$

$$x_1 = \quad F(y_1)$$

Numerical optimization methods may be used to minimise the functional J under the constraints (12.35). Methods, previously developed (linearization, Laplace transform, . . .), can also be adapted. A convolution approach may be used. We have:

$$y_1 = K * v$$
$$x_1 = F(y_1) \tag{12.36}$$

Substituting (12.36) in (12.34) gives a classical optimization problem whose solution is found by using a numerical method.

In conclusion, the aim of these approaches is to show the usefulness of drug interaction. Indeed, the interaction will be valuable if better results are obtained than with the drugs taken separately.

12.4 Numerical techniques for global optimization

Previously, we saw a global optimization technique called
Alienor which was used to calculate the global optimum of an
n-variable function.   It seems natural to think that other
approaches to global optimization [11] are possible, and
therefore the problem for applied mathematicians remains
wide open. Nevertheless, let us give two workable ideas:

a)   Elimination of variables

Let the optimization problem be:

$$\text{Min } [P(x_1, \ldots, x_n) = J] \qquad (12.37)$$

where P is a polynomial depending on $x_1, x_2, \ldots, x_n$.
Setting $y = P(x_1, \ldots, x_n)$, the problem (12.37) is
equivalent to:

Find the solution $(x_1^\dagger, \ldots, x_n^\dagger, y^\dagger)$ of:

$$P(x_1, \ldots, x_n) - y = 0$$

$$\partial P / \partial x_i = 0 \quad, \ i = 1, \ldots, n \qquad (12.38)$$

where $y^\dagger$ is the smallest possible value

Eliminating x_1 between $(P(x_1, \ldots, x_n) - y)$ and $\partial P / \partial x_1 = 0$
gives:

$$D_1(x_2, x_3, \ldots, x_n, y) = 0 \qquad (12.39)$$

Then x_1 is eliminated between $\partial P / \partial x_1$ and the other
derivatives, which gives a new algebraic system:

$$D_1(x_2, \ldots, x_n, y) \ = 0$$

$$R_{11}(x_2, x_3, \ldots, x_n) = 0$$

$$R_{12}(x_2, x_3, \ldots, x_n) = 0 \qquad (12.40)$$

$$\cdot$$
$$\cdot$$

$$R_{1(n-1)}(x_2, x_3, \ldots, x_n) = 0$$

Repeating the same process on (12.40) leads to:

$$D_2(x_3, x_4, \ldots, x_n, y) = 0$$

$$R_{21}(x_3, x_4, \ldots, x_n) = 0$$

$$\cdot \qquad\qquad\qquad\qquad (12.41)$$

$$\cdot$$

$$R_{2(n-2)}(x_3, x_4, \ldots, x_n) = 0$$

The process is continued until the abscissae $x_1, \ldots, x_n$ disappear. Finally a solution is obtained:

$$D_n(y) = 0 \qquad\qquad (12.42)$$

containing a single variable y.

The finding of the smallest solution of the polynomial $D_n(y) = 0$ remains. Note that $D_n(y)$ depends only on y, but it may be a higher degree polynomial. Nevertheless, classical methods, such as Bernouilli's technique, may be used to find the solution of $D_n(y) = 0$, numerically. The main difficulty is the elimination process which seems to be difficult to automatise on a computer. Notice that when $y^\dagger$ is obtained by solving (12.42), $x_n^\dagger$ is obtained from the solution of:

$$D_{n-1}(x_n, y^\dagger) = 0 \text{ and so on.} \qquad (12.43)$$

Thus we obtain successively $y^\dagger$, $x_n^\dagger$, $x_{n-1}^\dagger, \ldots, x_1^\dagger$. At each step it is only necessary to solve an algebraic (polynomial) equation depending on a single variable. Can this process be improved so that it is possible to use it on micro-computers?

Another idea for solving global optimization problems is as follows.

b)   Use of an approximation with separated variables

A very simple method for finding the minimum of $(x_1, \ldots, x_n)$ is to suppose that:

$$f(x_1, \ldots, x_n) = \sum_{i=1}^{n} f_i(x_i) \qquad (12.44)$$

The expression (12.44) is said to be a separated variable function. The minimum is easy to obtain. It is sufficient to notice that:

$$\underset{x_1,\ldots,x_n}{\text{Min}} \sum_{i=1}^{n} f_i(x_i) = \sum_{i=1}^{n} \underset{x_i}{\text{Min}} \ f_i(x_i) \qquad (12.45)$$

We are brought back to the minimization of functions f_i depending on a single variable x_i. Now a question arises! Is it possible to approximate a function $f(x_1, \ldots, x_n)$ by another function with separated variables? If it is possible, then the global optimization of f becomes very simple because we only need to find the global optimum of n functions depending on a single variable. Unfortunately, the answer is not obvious. From the numerical point of view a polynomial function may be defined:

$$P_m = \sum_{i,j} a_{ij} x_i^j \qquad (12.46)$$

and a functional can be introduced:

$$J = \int \ldots \int_0^T (P_m (x_1, \ldots, x_n) - f(x_1, \ldots, x_n))^2 dx_1 \ldots dx_n$$

$$(12.47)$$

If the minimum of J achieved is almost $J = 0$, then f may be approximated by a polynomial function with separated variables. The functional (12.47) can be replaced by a discretised functional involving some given points $(x_1^{(k)}, \ldots, x_n^{(k)})$ in R^n. The optimum of the continuous or discretised functional is calculate by solving the linear algebraic system:

$$\partial J / \partial a_{ij} = 0 \quad \text{for all } (i,j) \qquad (12.48)$$

A more general polynomial than (12.46) can be chosen. For example:

$$P_m = (x_1, \ldots, x_n) = a_0 + a_{11} u_1 + a_{12} u_1^2 + \ldots$$

$$+ a_{1m} u_1^m + \ldots + a_{n1} u_n + a_{n2} u_n^2 + \ldots$$

$$+ a_{nm} u_n^m \qquad (12.49)$$

where:

$$u_i = \sum_{i=1}^{n} \alpha_{ij} x_j \qquad , i = 1, \ldots, n \qquad (12.50)$$

In this case the coefficients a_{ij} and α_{ij} have to be identified, as before, by using an optimization method. However, in this case we do not have an algebraic linear system because the α_{ij} are also unknown. Nevertheless a linear technique may be used as for compartmental identification. Sequences of linear problems will be treated by fixing successively the α_{ij} and the a_{ij}. Similarly the approximation will be satisfactory if J is almost equal to zero at the optimum. When P_m is found, it will be minimised with respect to $u_1, \ldots, u_n$. The x_j result from the solution of an algebraic linear system (12.50).

The main inconvenience is the large number of parameters to be identified. Furthermore it is not proved that such an approximation is possible for all functions f.

12.5 Biofeedback and systems theory [32], [33], [34]

In some sense the retroactive systems may be considered as feedback systems, but the notion of bio-feedback is generally more specific. The word bio-feedback comes from cybernetics and describes the learning processes by which a subject (animal or human) can control a particular physiological variable (heart rate, arterial blood pressure, temperature, E.E.G., E.M.G, . . .) because of some information received about the state of the variable. This information is not generally provided by the organism itself, but instead comes from a device (often electronic) measuring the variable and transducing it to an audio or visual signal perceptible by the subject during the learning period.

During this period, the subject tries to control the signal. He is generally aware of its significance. By controlling the signal he controls the physiological variable. In some experimental procedures the subject is neither aware of, nor motivated by, the success of the performance [61]. Generally one can define a biofeedback system as a process of control of a physiological variable (internal) by means of an external signal depending on this variable. The subject may be considered as a system - a black box - receiving as input the stimulations corresponding to the experimental situation. The system's state is characterised by many variables, of which at least one is collected, measured and transformed into the input signal. Thus there is a closed loop, or feedback. An optimal control problem is set to the experimenter. He has to choose the retroaction signal (that is to say, the control law). Many biofeedback specialists strongly advocate use of the concepts and methods [34], [77] of systems theory. This approach allows the following:

1. Unification of the numerous interpretations of biofeedback (operant conditioning, learning of skills, visceral perception).

2. Deeper study of some aspects of the variable under control (stability, variability, response delay . .).

3. To direct a systematic choice of the nature of the biofeedback signal (binary, proportional, with delays . .). At least, this approach forces the researcher to establish a quantitative relation beteen the biofeedback signal and the subject's response. Then an optimal control calculation will give the best way of sending the information to the subject.

Only a few authors have been interested in this important aspect of biofeedback research, but some work is in progress. From an experiment on muscle learning [34], Gallego and colleagues proposed a first approach based on a linear stochastic learning model. At the same time a learning process associated with a respiratory biofeedback was studied and gave rise to some hypotheses, as follows.

- Prior to, and at the beginning of, the process, the measurements of the physiological variables studied are random.

- During the process the subject tries to fit his response to a desired value using a trial-and-error process to minimise the difference between this desired value and that of his responses.

- The probability of the fit increases with time, and the subject's responses tend towards the desired value.

These hypotheses are represented by the following model:

$$x_{n+1} = (x_n - \gamma F'(x_n))(1 - \epsilon_n) + \epsilon_n x_n^0$$

$$p_{n+1} = ap_n \quad \text{if} \quad \epsilon_n = 0 \ , \quad 0 < a < 1 \qquad (12.51)$$

$$p_{n+1} = p_n \quad \text{if} \quad \epsilon_n = 1$$

where:

- the sequence (x_n) corresponds to the observations of the controlled variable.

- F is the biofeedback signal as a function of the difference between x_n and the desired value.

- p_n is the probability of a fit.

- a, γ are parameters that have to be identified.

- x_n^0 is the value (at test n) of a random variable having the same statistical characteristics as the subject's responses before the learning process.

J. Gallego and colleagues [33] are interested in identifying the model parameters in real time. Much remains to be done, especially when considering non-linear learning processes. Furthermore, we do not have a general theory for studying the associated control problems.

12.6 Optimization of industrial processes

Many industrial processes are based on biochemical reactions and thus may be studied using methods developed in this book (modelling, optimization, optimal control). Numerous problems arise in the food industry, such that one may expect applied mathematicians to find a solution. Let us consider a practical example studied by J.P. Richard (E.N.S.I.A.), that of producing apple juice using an enzymatic method [12]. Some enzymes are able to break down the cellular wall of fruits and thus can replace mechanical pressing machines. To produce apple juice two succesive stages are required. In the first the temperature of the preparation is adjusted. In the second the enzymes can be added. Temperature and enzyme concentration have to be

chosen so that the viscosity is minimised at the end of the
second stage. A model relating the main parameters must be
prepared. As a first approach we can propose a simple
model:

$$V = F(T,C) \qquad\qquad (12.52)$$

which relates the viscosity V, the temperature T and the
enzyme concentration C. The function F has to be identified
from experimental data. Particular structures (polynomial,
exponential, . .) may be tried. Optimal control becomes a
classical minimization problem.

$$\underset{T,C}{\text{Min}}\ V^2 \qquad\qquad (12.53)$$

which can be solved numerically by the Alienor method.

     This modelling does not explicitly involve time. More
sophisticated models may be necessary in view of numerical
tests and experimental results. For example, differential
equations may replace (12.52).

12.7 Optimality in physiology

To   ensure   uniqueness in compartmental   modelling   an
optimization criterion was introduced. This criterion was
based on an optimality hypothesis. Later we developed a
thermo-regulation model in which a minimization criterion
was used. It is fairly certain that many physiological
systems can be studied by introducing a minimization
criterion related to the energy consumed by the system. At
the microscopic (cellular) level this minimization   is
probably often false but one can expect it to be valid at
the macroscopic level. Of course, physiological systems
must be considered as belonging to the "macroscopic" set of
biological systems. After modelling the phenomenon
according to the available data (physical laws, experimental
results) the main difficulty is in defining a mathematical
expression for the system's energy. Adding to the model the
requirement for minimization of this relation should give
numerical results agreeing with the physiological data.   In
many cases the addition of a criterion will ensure
uniqueness when solving the mathematical model describing
the phenomenon [7]. The reader has to choose his own system
according to his ability and preference.

CHAPTER 13

CONCLUSIONS

It is obvious that it is not possible to describe all the tools [66] used for modelling biological systems in one book. We have given some (perhaps the most important) but have not, for example, developed data analysis, which is a very important technique for determining the most significant variables or parameters in a biological problem [26]. In this book, our aims were as follows:

- to give the main techniques for modelling biological phenomena;

- to study some concrete problems as a means of describing the main difficulties in modelling;

- to describe the fundamental numerical methods used when identifying a model, or controlling a biological system;

- to show that many problems remain open, and that co-operation between mathematicians and biologists could help in their solution;

- to prove that a mathematical and numerical study of a medical or biological phenomenon can always improve the knowledge of the system;

- to assert that complex systems involving regulation (as for instance optimised processes) can only be studied by using models and mathematical methods [19], [50], [55], [63].

Nevertheless, it is not easy to become and remain a biomathematician because of the hostility of many mono-disciplinary researchers. Pure mathematicians do not like concrete applications very much, and consequently often consider biomathematics as uninteresting or trivial mathematics. They do not understand the difficulties of modelling, nor the usefulness of appropriate numerical methods (for identification, optimization . .). On the other hand, biologists and physicians are not always convinced of the value of using mathematical and numerical techniques in their disciplines. They think they can treat the problems by themselves. Thus some good examples of successful collaboration have to be given as a proof of the utility. When new properties are proved, or new possibilities (optimization) are demonstrated by mathematical modelling, then its usefulness is beyond doubt. For instance, optimal effects of drugs cannot be defined without using modelling and optimization methods.

Some important statistical problems are not studied in this book. Statistical variability within, or between individuals was not taken into account. The main reason for the exclusion is that there is usually insufficient data from biological or medical experiments to allow statistical treatment. The second (less important!) reason is that this author is not as competent as some others [26] to treat this type of problem. However, it must be observed that of the concrete biological problem set to our laboratory, none allowed sophisticated statistical treatment.

Our main objective was thus to propose an appropriate mathematical technique of modelling taking into account the numerical data and the physical properties [66]. With this it becomes possible to act on the biological system by means of control variables. From my point of view, this possibility of acting on a biological system is the most interesting property of biomathematical models.

THE ALIENOR PROGRAM (IMPLEMENTED BY A. GUILLEZ)

As already mentioned, this method uses a transformation to reduce an n-variable function to a single-variable one. The approximation may be to any desired accuracy. Once the transformation has been defined it is easy to find every extremum of an n-variable function. Note that the transformation has a tree structure (see Figure A.1 below):

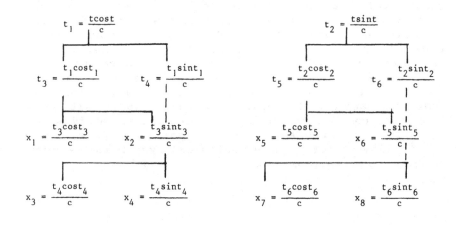

(FIG. A1)

A test program (see below) can be written to evaluate the maximum distance between a point of R^n and the Alienor curve. More precisely, let L be a point of R^n, and then the Alienor curve is at distance LM from this point L, where M is the image of L in the transformation. Using the densification technique (see below) and the numerical values given in the following table, a relative distance may be

evaluated by a statistical technique (see test program).
The relative distance (with OL = distance from L to the
origin) is:

LM/OL = 0.03 with a dispersion equal to 1, that is, a
standard deviation also equal to 0.03, which implies that
the probability of obtaining LM/OL $\geq$ 0.09 is 5%.

For evaluating this maximum distance there is no
possibility of a mathematical proof. We only have a
numerical proof based on a statistical technique of choosing
random points.

No. of transformed variables	Divisor c	Levels s	θ_{max}	$c-2\pi$	$e^{s^2}(c-2\pi)$
2	30	1	40	24	65.23
3 to 4	7	2	60	0.72	39.31
5 to 8	2π	3	260	0	
9 to 16	2π	4	1586	0	

Multiplier K: $2 \leq K \leq 3$

where s is the number of levels in the tree, and θ
represents the Alienor variable after transformation (θ_{max}
is the maximum value of θ).   The multiplier K is useful for
defining the exploration step.

Table A.1

The Alienor technique is simple, but it is difficult to
reduce the calculation time. Because of this, θ_{max} has to
be limited as in Table A.1. Another idea for reducing the
calculation time is to densify the numerical exploration.
We have a lower and an upper bound (m_i and M_i) for each
variable x_i which was transformed. Denoting by ζ_i the
Alienor function corresponding to x_i we set:

if $\zeta_i \in [0,1]$ then $x_i = m_i + (M_i - m_i)\zeta_i$

if $\zeta_i \notin [0,1]$ then $x_i = m_i + (M_i - m_i)(K\zeta_i - INT(K\zeta_i))$

Practical rules have been defined numerically by A. Guillez. For example:

$$\theta_M = INT(c^s + 10.5)$$

$$s = 1 + INT(Log(N - 0.8)/Log\ 2) \quad \text{for all N}$$

$$c = 2\pi + exp(-s^{2.17}).65$$

giving, for $s = 1$, $c = 30.28$

for $s = 2$, $c = 7.005$

for $s = 3$, $c = 6.2844$ and so on.

On microcomputers (such as Apple II or Commodore 64), problems were solved having numbers of variables up to 16. The following program is concerned with the identification of a_i and λ_i appearing in:

$$x(t) = \sum_{i=1}^{n} a_i\ exp(-\lambda_i t)$$

where $x(t)$ is measured at $t = t_j$, $j = 1, \ldots, m$.

The program has been tested on concrete examples from pharmacokinetics and on random functions $x(t)$.

TEST PROGRAM

MAX:DS= 7.28000001 %MEAN:DI= 2.43354 %MIN:DM= .117 %C= 7

K= 2.5

```
DL( 0 )= .499
DL( 1 )= .283
DL( 2 )= 2.777
DL( 3 )= 1.456
DL( 4 )= 2.13
DL( 5 )= 1.099
DL( 6 )= 1.132
DL( 7 )= 7.28000001
DL( 8 )= .712
DL( 9 )= 3.888
DL( 10 )= 1.046
DL( 11 )= 2.469
DL( 12 )= 2.167
DL( 13 )= .117
DL( 14 )= .554
DL( 15 )= 1.541
DL( 16 )= 3.49
DL( 17 )= 1.028
DL( 18 )= 1.181
DL( 19 )= 2.497
DL( 20 )= 1.763
ALIENORTEST GUILLEZ 30/5/84
```

```
1    INPUT C0,K:H=1/K:PA=.2:CR=0:TM=C0↑2 +10
2    DIM DL(20)
3    L(0)=RND(8)
4    FOR I=1 TO 3:L(I)=L(I-1)+(1-L(I-1))*RND(8):NEXT I
5    SL=0
6    FOR I=0 TO 3:SL=SL+L(I)↑2:NEXT I
7    IF SL<1 THEN 3
8    PRINT "START":C=C0
10   TA=TM*RND(8):T=TA:GOSUB 400:DA=DE
11   FOR II=1 TO 12:T=TM*RND(8):GOSUB 400
12   IF DE<DA THEN TA=T:DA=DE
13   NEXT II
14   GOSUB 300:DR=DL:TR=TL:TC=TL:PRINT "DR="DR
20   DZ=DR:TZ=.2
22   Q=0
24   TA=TZ+PA*Q:T=TA:GOSUB 400:DA=DE
26   IF T>TM THEN 38
28   IF DA>DZ THEN Q=Q+1:GOTO 24
```

```
30   GOSUB 300:DZ=DL:TZ=TL:TC=TL:PRINT "DZ="DZ
32   IF TL<(TA+.2) THEN TZ=TA+.2
34   GOTO 22
38   TA=TC:DA=DZ:GOSUB 300:DL(CR)=DL:PRINT "DL("CR")="DL
40   IF CR>=20 THEN 44
42   CR=CR+1:GOTO 3
44   DS=DL(0):DM=DL(0):DI=DL(0)↑2
46   FOR J=1 TO 20:DI=DI+DL(J)↑2
50   IF DL(J)>=DS THEN DS=DL(J)
52   IF DL(J)<=DM THEN DM=DL(J)
54   NEXT J
56   DI=1E-5*INT(1E5*SQR(DI/21)+.5):PRINT "DS="DS "DM="DM
                                                         "DI="DI
58   OPEN 3,4
60   CMD 3
61   PRINT "MAX:DS="DS"%" "MEAN:DI="DI"%" "MIN:DM="DM"%"
                                                     "C="C "K="K
62   FOR I=0 TO 20:PRINT "DL("I")="DL(I):NEXT I
63   PRINT "ALIENORTEST GUILLEZ 30/5/84"
65   PRINT #3
66   CLOSE 3
70   REM EXAMPLE OF PROGRAM ALIENORTEST 4 VARIABLES LESS THAN 1
72   REM 21 PTS L RND, 21 CORRESPONDING POINTS M ON ALIENORWAY
74   REM "400" COMPUTES DE=100*LM/OM
76   REM AND FOR EACH PT L 300 COMPUTES DL=INF(DE)
78   REM DS=SUP(DL(CR)) DI=MEAN(DL(CR)) DM=INF(DL(CR))
80   END
300  P=DA*.002:X(0)=TA:Y(0)=DA
302  IF P>PA THEN P=PA
304  X(1)=X(0)+P:T=X(1):GOSUB 400:Y(1)=DE
306  IF Y(1)<Y(0) THEN X(0)=X(1):Y(0)=Y(1):GOTO 304
308  X(2)=X(0):Y(2)=Y(0)
310  X(3)=X(2)-P:T=X(3):GOSUB 400:Y(3)=DE
312  IF Y(3)<Y(2) THEN X(2)=X(3):Y(2)=Y(3):GOTO 310
314  X(4)=X(0)+P*Y(0)/(Y(0)+Y(1)):T=X(4):GOSUB 400:Y(4)=DE
316  X(5)=X(2)-P*Y(2)/(Y(2)+Y(3)):T=X(5):GOSUB 400:Y(5)=DE
318  TL=X(0):DL=Y(0)
320  FOR N=1 TO5
322  IF Y(N)<=DL THEN TL=X(N):DL=Y(N):N1=N
324  NEXT N
326  IF P<.002 THEN 332
328  X(0)=TL:Y(0)=DL:P=P*.55
330  GOTO 304
332  RETURN
400  T0=T*COS(T)/C:T1=T*SIN(T)/C
```

```
402 T(0)=T0*COS(T0)/C:T(1)=T0*SIN(T0)/C:T(2)=T1*COS(T1)/C:
                                        T(3)=T1*SIN(T1)/C
404 IF T(0)>=0 AND T(0)<=H THEN M(0)=K*T(0):GOTO 408
406 M(0)=K*T(0)-INT(K*T(0))
408 FOR I=1 TO 3
410 IF T(I)>=0 AND T(I)<=H THEN
                        M(I)=M(I-1)+(1-M(I-1))*K*T(I):GOTO 414
412 M(I)=M(I-1)+(1-M(I-1))*(K*T(I)-INT(K*T(I)))
414 NEXT I
420 DE=0:SE=0
422 FOR I=0 TO 3:SE=SE+ABS(M(I)):DE=DE+(M(I)-L(I))↑2
424 NEXT I
425 IF SE=0 THEN DE=1E4:GOTO 430
426 DE=1E-5*INT(1E5*SQR(DE)/SE+.5)
428 DE=100*DE
430 RETURN
```

ALIENOR METHOD FOR IDENTIFICATION OF AN EXPONENTIAL DEVELOPMENT

```
2   DIM T(11),C(11),CC(11),MC(11,3),CM(3,11),M(3,3),V(3),L(3)
4   DIM LL(3),A(3),AL(3),LX(3),P(3),X(5),Y(5),G(3)
6   DATA .5,1,2,4,6,8,11,15,18,21,24,30
8   FOR I=0 TO 11:READ T(I):NEXT
10  L(0)=.30*RND(8):L(1)=RND(8):L(2)=L(1)+(1-L(1))*RND(8):
                                          L(3)=4+2*RND(8)
12  A(0)=2+4*RND(8):A(1)=-2-2*RND(8):A(2)=.5-5*RND(8):
                                          A(3)=3*5RND(8)
14  I=0:C(I)=0
16  J=0:SI=INT(RND(8)*100)
18  C(I)=C(I)+EXP(-T(I)*L(J))*A(J)
20  IF J<=2 THEN J=J+1: GOTO 18
22  IF C(I)<=0 THEN 12
23  C(I)=C(I)*(1+.05*RND(8)*(-1)↑SI):C(I)=.001*INT(C(I)*1000+.5)
24  IF I<=10 THEN I=I+1:GOTO 16
25  OPEN 3,4
26  CMD 3
27  PRINT "GIVEN"
28  FOR I=0 TO 9 STEP 3
29  PRINT "C("I")="C(I)"*""C("I+1")="C(I+1)"*""C("I+2")="C(I+2)
30  NEXT I
31  PRINT "************************"
32  PRINT #3
33  CLOSE 3
34  T=60*RND(8):GOSUB 400:DA=DE:TA=T
36  FOR II=0 TO15:T=50*RND(8):GOSUB 400
38  IF DE<=DA THEN DA=DE:TA=T
40  NEXT II
42  GOSUB 300:DR=DM:TR=TM:PRINT "DR="DR
44  TZ=0:DZ=DR
46  Q=0
48  T=TZ+.2*Q:GOSUB 400:DA=DE:TA=T
50  IF T>=60 THEN 60
52  IF DA>DZ THEN Q=Q+1:GOTO 48
54  GOSUB 300:DZ=DM:TZ=TM:TC=TM
56  IF TZ<=(TA+.2) THEN TZ=TA+.2
58  GOTO 46
60  IF TZ=0 THEN TC=TR
62  TA=TC:DA=DZ:GOSUB 300:DL=DM
64  OPEN 3,4
66  CMD 3
```

```
68    PRINT "RESULTS FROM ALIENOR"
70    FOR II=0 TO 3:LL(II)=L(II):AL(II)=A(II)
72    PPRINT "LL("II")="LL(II) "AL("II")="AL(II)
74    NEXT II
75    FOR M=0 TO 11 STEP 3
76    PRINT "CC("M")="CC(M)"*""CC(M+1")="CC(M+1)"*""CC("M+2")="
                                                    CC(M+2)
77    NEXT M
78    PRINT "DL="DL
80    PRINT "*************************"
82    PRINT #3
84    CLOSE 3
86    GOSUB 200
88    OPEN 3,4
90    CMD 3
92    PRINT "TRUE VALUES"
94    PRINT "DM="DM
96    FOR II=0 TO 3:PRINT"L("II")="L(II) "A("II")=A(II):NEXT II
97    FOR II=0 TO 11 STEP 3
98    PRINT "CC("II")="CC(II)"*""CC("II+1")="CC(II+1)
                                    "*""CC("II+2")="CC(I+2)
99    NEXT II
100   PRINT #3
102   CLOSE 3
104   END
200   RG=0:DX=DL
202   FOR I2=0 TO 3:LX(I2)=LL(I2):P(I2)=LL(I2)*DL/100:NEXT I2
204   FOR I2=0 TO 3:Y(0)=LX(I2):X(0)=DX
206   Y(1)=Y(0)+P(I2):L(I2)=Y(1):GOSUB 420:X(1)=DE
208   IF X(1)<X(0) THEN Y(0)=Y(1):X(0)=X(1):GOTO 206
210   Y(2)=Y(0):X(2)=X(0)
212   Y(3)=Y(2)-P(I2):L(I2)=Y(3):GOSUB 420:X(3)=DE
214   IF X(3)<X(2) THEN Y(2)=Y(3):X(2)=X(3):GOTO 212
216   Y(4)=Y(0)+P(I2)*X(0)/(X(0)+X(1)):L(I2)=Y(4):GOSUB 420:
                                                    X(4)=DE
218   Y(5)=Y(2)-P(I2)*X(2)/(X(2)+X(3)):L(I2)=Y(5):GOSUB 420:
                                                    X(5)=DE
220   DX=X(0):LX(I2)=Y(0)
222   FOR J2=1 TO 5
224   IF X(J2)<=DX THEN DX=X(J2):LX(I2)=Y(J2)
226   NEXT J2
228   NEXT I2
229   RG=P(0)↑2+P(1)↑2+P(2)↑2+P(3)↑2
230   IF SQR(RG)<=.004 THEN 236
232   FOR I2=0 TO 3:P(I2)=.55*P(I2):NEXT I2
```

```
234 GOTO 204
236 L(0)=LX(0):L(1)=LX(1):L(2)=LX(2):L(3)=LX(3)
238 GOSUB 420:DM=DE
240 RETURN
300 P=DA*.002:X(0)=TA:Y(0)=DA:RF=0
301 IF P>.2 THEN P=.2
302 X(1)=X(0)+P:T=X(1):GOSUB 400:Y(1)=DE
304 IF Y(1)<=(Y(0)-1E-5) THEN X(0)=X(1):Y(0)=Y(1) GOTO 302
306 X(2)=X(0):Y(2)=Y(0)
308 X(3)=X(2)-P:T=X(3):GOSUB 400:Y(3)=DE
310 IF Y(3)<=(Y(2)-1E-5) THEN X(2)=X(3):Y(2)=Y(3) GOTO 308
312 X(4)=X(0)+P*Y(0)/(Y(0)+Y(1)):T=X(4):GOSUB 400:Y(4)=DE
314 X(5)=X(2)-P*Y(2)/(Y(3)+Y(2)):T=X(5):GOSUB 400:Y(5)=DE
316 TM=X(0):DM=Y(0)
318 FOR I1=1 TO 5
320 IF Y(I1)<Y(0) THEN TM=X(I1):DM=Y(I1)
322 NEXT I1
324 IF P<=.002 THEN 330
326 X(0)=TM:Y(0)=DM:P=.55*P
328 RF=RF+1:GOTO 302
330 T=TM:GOSUB 400
332 RETURN
400 T0=T*COS(T)/7:T1=T*SIN(T)/7
402 T2=T0*COS(T0)/7:T3=T0*SIN(T0)/7:T4=T1*COS(T1)/7:
                                        T5=T1*SIN(T1)/7
404 IF T2>=0 AND T2<=.3 THEN L(0)=T2:GOTO 408
406 L(0)=T2-.3*INT(3.3333*T2)
408 IF T3>=0 AND T3<=.3334 THEN L(1)=3*T3:GOTO 412
410 L(1)=3*T3-INT(3*T3)
412 IF T4>=0 AND T4<=.3334 THEN L(2)=L(1)+(1-L(1))*3*T4:
                                        GOTO 416
414 L(2)=L(1)+(1-L(1))*(3*T4-INT(3*T4))
416 IF T5>=0 AND T5<=.3334 THEN L(3)=6*T5+4:GOTO 420
418 L(3)=4+6*T5-2*INT(3*T5)
420 DE=0:SE=0
422 FOR I=0 TO 3:V(I)=0:M(I,0)=0:M(I,1)=0:M(I,2)=0:M(I,3)=0
424 L(I)=ABS(L(I)):L(I)=INT(L(I)*1E4+.5)*1E-4
426 NEXT I
434 FOR I=0 TO 11:FOR J=0 TO 3
436 E=EXP(-T(I)*L(J)):MC(I,J)=E:CM(J,I)=E
438 NEXT J,I
440 FOR I=0 TO 3:FOR J=0 TO 11
442 V(I)=V(I)+CM(I,J)*C(J)
444 FOR K=0 TO 3:M(I,K)=M(I,K)+CM(I,J)*MC(J,K):NEXT K
446 NEXT J,I
```

```
448 FOR H=3 TO 1 STEP -1:FOR I=H-1 TO 0 STEP -1
450 IF M(H,H)<>0 THEN 470
452 FOR L=H-1 TO 0 STEP -1
454 IF M(L,H)<>0 THEN 460:NEXT L
456 DE=10000:GOTO 500
460 REM ALIENOR
462 FOR W=0 TO 3
464 G(W)=M(L,W):M(L,W)=M(H,W):M(H,W)=G(W)
466 NEXT W
468 V1=V(L):V(L)=V(H):V(H)=V1
470 V(I)=V(I)-M(I,H)*V(H)/M(H,H)
472 FOR J=0 TO H
474 M(I,J)=M(I,J)-M(I,H)*M(H,J)/M(H,H)
476 NEXT J,I,H
478 A(0)=V(0)/M(0,0)
479 A(0)=INT(A(0)*1E4+.5)*1E-4
480 FOR I=1 TO 3:S=0
482 FOR K=0 TO I-1:S=S+M(I,K)*A(K):NEXT K
484 A(I)=(V(I)-S)/M(I,I)
485 A(I)=INT(A(I)*1E4+.5)*1E-4
486 NEXT I
488 FOR I=0 TO 11:CC(I)=0
490 FOR J=0 TO 3:CC(I)=CC(I)+A(J)*MC(I,J):NEXT J
492 CC(I)=INT(CC(I)*1000+.5)*.001
494 SE=SE+ABS(CC(I)):DE=DE+(CC(I)-C(I))↑2
496 NEXT I
498 DE=100*SQR(12*DE)/SE
499 DE=INT(DE*1E5+.5)*1E-5
500 RETURN
```

EXAMPLE OF OUTPUT

GIVEN

```
C(0) = 0.551 * C(1)  = 0.714 * C(2)  = 1.500
C(3) = 1.772 * C(4)  = 1.665 * C(5)  = 1.461
C(6) = 1.140 * C(7)  = 0.885 * C(8)  = 0.691
C(9) = 0.546 * C(10) = 0.430 * C(11) = 0.268
************************
```

RESULTS FROM ALIENOR

```
LL(0) = 0.0776    AL(0)  =  2.7817
LL(1) = 0.3547    AL(1)  = -0.7009
LL(2) = 0.9532    AL(2)  = -3.6703
LL(3) = 5.426     AL(3)  = 11.1678
CC(0) = 0.551  * CC(1)  =  0.717  * CC(2)  = 1.492
CC(3) = 1.789  * CC(4)  =  1.651  * CC(5)  = 1.452
CC(6) = 1.17   * CC(7)  =  0.865  * CC(8)  = 0.687
CC(9) = 0.545  * CC(10) =  0.432  * CC(11) = 0.271
DL    = 1.32261
************************
```

TRUE VALUES

```
DM    = 1.30095
L (0) = 0.0775    A(0)  =  2.7837
L (1) = 0.3524    A(1)  = -0.746
L (2) = 0.981     A(2)  = -3.6999
L (3) = 5.4233    A(3)  = 11.5018
CC(0) = 0.551  * CC(1)  =  0.715  * CC(2)  = 1.495
CC(3) = 1.786  * CC(4)  =  1.648  * CC(5)  = 1.452
CC(6) = 1.171  * CC(7)  =  0.867  * CC(8)  = 0.689
CC(9) = 0.546  * CC(10) =  0.433  * CC(11) = 0.272
```

REFERENCES

1. P.M. Anselone: Non-linear integral equations, Madison Wisconsin Press (1964).

2. C.T.H. Baker and C. Phillips: The numerical solution of non-linear problems, Clarendon Press (1981).

3. S. Barnett: Introduction to mathematical control theory, Clarendon Press (1975).

3a. R. Bellman: Introduction to matrix analysis, McGraw-Hill (1960).

4. R. Bellman: Mathematical methods in medicine, World Scientific (1983).

5. R. Bellman and R. Kalaba: Dynamic programming and modern control theory, Academic Press (1965).

6. R.B. Bird, W.E. Stewart and E. Lightfoot: Transport phenomena, Wiley (1960).

7. D. Boichut and F. Valentini: Human locomotion analysis. Determination of muscular forces and nervous orders. Int. J. Bio Medical Computing, 14, pp. 217-230 (1983).

8. V.W. Bolie: Coefficients of blood glucose regulation. J. Applied Physiology, 16, pp. 783-788 (1961).

9. A. Caron: Méthode de résolution numériques d'équations intégrales non linéaires à noyaux réguliers par un schéma itératif sous contrainte de régularité. C.R.A.S. 13, p. 649 (1981).

10. A. Caron: Méthodes numériques pour la résolution d'équations intégrales linéaires at non linéaires. Contribution à la définition et à l'approximation des parties finies. Thèse de 3ème cycle, Paris (June 1981).

11. Y. Cherruault: Biomathématiques, collection 'Que sais-je?' Presses Universitaires de France (1983).

12. Y. Cherruault et Coll: Recherches biomathématiques - Méthodes et exemples, CIMPA Ed. OFFILIB, Diffusion (1983).

13. Y. Cherruault and M. Guerret: Parameters identification and optimal control theory in pharmacokinetics. Acta Applicandae Mathematicae, 1, pp. 105-120 (1983).

14. Y. Cherruault and A. Guillez: A simple method for optimization. Kybernetes (1983).

15. Y. Cherruault and A. Guillez: L'obtention de tous les minima des fonctions de plusieurs variables et du minimum absolu: Aliénor. Electronique, Techniques et

Industries, (to appear 1985).

16. Y. Cherruault and A. Guillez: Quelques méthodes
     numériques pour la résolution d'équations intégrales
     non linéaires. Note to C.R.A.S. (1980).

17. Y. Cherruault and A. Guillez: Méthodes numériques pour
     la résolution d'équations intégrales non linéaires.
     Calcolo, vol. XVIII fas. IV, pp. 385-407 (1981).

18. Y. Cherruault and P. Loridan: Modélisation et méthodes
     mathématiques en biomédicine, Masson, Paris (1977).

19. C. Chevalet and A. Micali: Modèles mathématiques en
     biologie, Springer Verlag (1981).

20. P.G. Ciarlet: Introduction à l'analyse numérique
     materielle et l'optimisation, Masson, Paris (1982).

21. D. Claude and B. Weil: Découplage et immersion d'un
     modèle neuro-endocrinien. Note to C.R.A.S. (1984).

22. R. Comolet: Biomécanique circulatoire, Masson, Paris
     (1984).

23. E.T. Copson: Partial differential equations, Cambridge
     University Press (1975).

24. M. Crouzeix and A. Mignot: Analyse numérique des
     equations différentielles, Masson, Paris (1984).

25. P. J. Davies: Interpolation and approximation,
     Blaisdell (1965).

26. P. Deheuvels: How to use bio-equivalence studies? The
     right use of confidence intervals. J. of organizational
     behaviour and statistics, 1, no. 1, pp. 1-15 (1984).

27. P. Delattre: L'évolution des systèmes moléculaires,
     Maloine (1971).

28. P. Delattre: Système, structure, fonction, évolution,
     Maloine (1984).

29. J. Delforge: Etude sur les problèmes d'identification
     dans les systèmes de transformations linéaires. Thèse
     de 3ème cycle, Paris 6 (February 1975).

30. L. Dorveaux: Un modèle mathématique d'échanges
     thermiques trans-cutanés. Optimisation de l'énergie de
     transfert. Thèse de 3ème cycle, Paris 6 (October 1983).

31. M. Dupuy and A. Guillez: Diffusion et osmoses
     selectives. Application au problème de fonctionnement
     du nephron. Int. J. of Bio Medical Computing, 10,
     (1979).

32. J. Gallego and L. Laurenti-Lions: Modèle mathématique
     d'un processus d'apprentissage par biofeedback. Int. J.
     of Bio Medical Computing, 13, (1982).

33. J. Gallego and L. Laurenti-Lions: A systemic model of
     biofeedback learning. Kybernetes, 12, (1983).

34. J. Gallego, L. Laurenti-Lions, B. Chambille, G. Vardon and C. Jacquemin: Conditionnement opérant myoeléctrique chez des sujets ignorant le but de l'expérience. C.R.A.S., Paris 1982).

35. M. Gibaldi and D. Perrier: Pharmocokinetics, M. Dekker (1975).

36. K. Godfrey: Compartmental models and their application, Academic Press (1983).

37. A. Guillez and Y. Cherruault: Identification des systèmes compartimentaux ordinaires. Les programmes FELICITY et SOPHIE-ALICE. Proceedings of the 10th Congrès Int. de Cybernetique, Namur (1983).

38. P. Henrici: Discrete variable methods in ordinary differential equations, Wiley (1961).

39. S. Iyengar: Computer modeling of complex biological systems, C.R.C press (1984).

40. J.A. Jacquez: Compartmental analysis in biology and medicine, Elsevier (1972).

41. S. Jaggi and A. Caron: The solution of an integral equation with its application to human joints. Int. J. of Bio Medical Computing, 14, pp. 249-262 (1983).

42. D.S. Jones and B.D. Sleeman: Differential equations in mathematical biology, G. Allen and Unwin (1983).

43. Ch. Kaiser: Physiologie 3. Les grandes fonctions, Flammarion (1970).

44. R. Kalaba and K. Spingarn: Control, identification and input optimization, Plenum Press (1982).

45. S. Kamga: Optimisation du traitement des diabétiques. Thesis, University Paris 6. (To appear).

46. V. Karmanov: Programmation mathématique, MIR, Moscow (1977).

47. I. Karpouzas: Etude de la mécanique ventilatoire par application du principe de la minimisation de l'énergie, par des méthodes mathématiques et numériques. Thesis, Paris 6 (November 1983).

48. J.P. Kernevez: Evolution et contrôle de systèmes biomathématiques. Thèse d'Etat, Paris (March 1972).

49. G.I. Marchuk: Méthodes de calcul numérique, MIR, Moscow (1980).

50. G.I. Marchuk and I.N. Belykh: Mathematical modeling in immunology and medicine, North Holland (1983).

51. S. Metry: Côntrole des grandeurs hemodynamiques de la circulation sanguine. Thèse de 3ème cycle, Paris 6 (1981).

52. R.K. Miller: Non linear Volterra integral equations, Benjamin (1971).

53. P. Nelson: Logique du neurone et du système nerveux, Maloine (1978)

54. R. Ozon: Regulation of Xenopus oocyte meiotic maturation. Hormones and Cell Regulation, 7, pp. 287-298 (1983).

55. J.E. Randall: Microcomputers and Physiological simulation, Addison-Wesley.

56. P.A. Raviart and J.M. Thomas: Introduction à l'analyse numérique des équations aux dérivées partielles, Masson, Paris (1983).

57. R.D. Richtmayer: Difference methods for initial value problems, Wiley Interscience (1962).

58. F. Santi: Identification du modèle de régulation des couples ago-agonistes dans le cadre du système endocrinien surréno-posthypophysaire. Côntrole-Thèse de 3ème cycle, Paris 6 (June 1982).

59. L. Schwartz: Théorie des distributions, Hermann (1966).

60. L. Schwartz: Méthodes mathématiques pour les sciences physiques, Hermann (1961).

61. G.E. Schwartz and J. Beatti: Biofeedback - Theory and Research, Academic Press (1977).

62. B. Some: Identification, côntrole optimal et optimisation dans les systèmes d'équations différentielles compartimentales. Thèse de 3ème cycle, Paris 6 (June 1984).

63. J.D. Spain: Basic microcomputer models in biology, Addison-Wesley (1982).

64. G.W. Swan: An optimal control model of diabetes mellitus. Bull. of Math. Biology, 44, no. 6, pp. 793-808 (1982).

65. G.W. Swan: Applications of optimal control theory in biomedicine, M. Dekker (1984).

66. R. Thom: Stabilité structurelle et morphogénèse, Benjamin (1972).

67. D. Thomas and J.P. Kernevez: Analysis and control of immobilized enzyme systems, North-Holland (1976).

68. N. Tosaka and S. Mikaye: An analysis of a linear diffusion problem with Michaelis-Menten kinetics by an integral equation method. Bull. of Math. Biology, pp. 841-845 (1982).

69. F.G. Tricomi: Integral Equations, Wiley Interscience (1963).

70. J. Vignes: Méthodes numeriques d'optimisation d'une
    fonction de plusieurs variables, Chimie Industrie -
    génie chimique, 97, no.8, pp. 1264-1276 (1967).

71. J. Vignes: New methods for evaluating the validity of
    results of mathematical computations. Mathematics and
    computers in simulation, vol. XX, no. 4, pp. 227-249
    (1978).

72. J. Wagner: Fundamentals of clinical pharmcokinetics,
    Drug Intelligence (1975).

73. E. Walter: Identifiability of state space models.
    Lecture notes in biomathematics, Springer Verlag
    (1982).

74. E.B. Weil: Formalisation et contrôle du système
    endocrinien surréno-posthypophysaire par le modèle
    mathématique de la régulation des couples ago-
    antagonistes. Thèse d'Etat, Paris 6 (1979).

75. S. Wright: Physiologie appliquée à la Médicine,
    Flammarion (1972).

76. R.A. Willoughby: Stiff differential equations, Plenum
    (1974).

77. A.J. Yates: Biofeedback and modification of behaviour.
    Plenum (1980).

INDEX